PRAISE FOR STEVE LOHR'S *GO TO*

"*Go To* is smooth, creamy, and sometimes delightful. It is a sort of *Iliad* for the computer age, the epochal story of how the 'software revolution' came about and who did what. The story never bogs down because he tells it as a series of miniature biographies. . . . Mr. Lohr writes fine, translucent prose, engaging and never overwrought. And he knows the field well."

—David Gelernter, *New York Times*

"*Go To* is an enlightening read and does a fine job of demonstrating the power of imagination. If you can imagine it and code it, you can indeed change the world."

—*Boston Sunday Globe*

". . . a clear, understandable introduction to a host of thorny technical concepts. An excellent primer for anyone curious about the insides of a PC, *Go To* is also Lohr's reply to John McCarthy, one of the great gray-beards of computer science, who complained to Lohr about 'ignorant journalists.' . . . Lohr's book should be required reading for any journalist who covers the field."

—*New York Times Books Review*

"This is no textbook. It is history; it is not technical, and it is told through the amazing personalities who created the programming languages and the software that make computers do their tricks. . . . [*Go To*] is clear, no nonsense journalism. The programmers we meet are captured by the thrill of being engineers who work without building materials and saws and hammers. They can build something out of nothing, and limited only by their imagination."

—*International Herald Tribune*

"If Steve Lohr were a programmer, his code would be coherent and well ordered, proceeding to the proper subroutines without baroque diversions or mind-jarring shortcuts. That's the way he writes: with clear prose that makes sense of a complicated subject in *Go To*, a whirlwind meet-and-greet of 'math majors, bridge players, chess wizards, maverick scientists and iconoclasts. . . the programmers who created the software revolution.'"

—*Newsweek*

Go To

Also by Steve Lohr

U.S. v. Microsoft: The Inside Story of the Landmark Case
(co-authored with Joel Brinkley)

THE STORY OF THE MATH MAJORS,
BRIDGE PLAYERS, ENGINEERS,
CHESS WIZARDS, MAVERICK
SCIENTISTS AND ICONOCLASTS—
**THE PROGRAMMERS WHO
CREATED THE SOFTWARE
REVOLUTION**

Go To

STEVE LOHR

BASIC
BOOKS

A Member of the Perseus Books Group

Copyright © 2001 by Steve Lohr
Published by Basic Books,
A Member of the Perseus Books Group

Designed by Bookcomp, Inc.

A CIP catalog record for this book is available from the Library of Congress.
ISBN 0-465-04226-0

02 03 04 / 10 9 8 7 6 5 4 3 2 1

To Fred, Terry and Nikki

Contents

Acknowledgments

A seed for this book was planted when Jim Gray passed through New York in the spring of 1999 on his way to pick up his Turing Award, an accolade that has been called the Nobel prize of computer science. A California native, Gray was educated at Berkeley in the 1960s and has worked nearly all his life in Silicon Valley. Gray is a veteran of many a business cycle in the Valley, and he shook his head with bemused disdain at the Internet investment mania of the time – all elevator pitches and IPOs. People seemed to be a lot more excited about money than about technology – a world askew, by Gray's standards.

Sure, he said, people with skills make a good living in the field. "But it's not about money," Gray observed. "The joy and the real appeal is to be able to play and build things with this cool technology of software."

I thought it might be fun to take a deeper look at the history of computer programming, and talk to the architects and builders of the software world that we increasingly live in. An added benefit was that because computing has moved so rapidly – and is such a comparatively recent field – most of the pioneers of programming are still alive.

It seemed like a good idea at the time.

Some others thought it might be an intriguing project as well, and they deserve my thanks. Without the encouragement of the agents at Brockman Inc. – John Brockman and Katinka Matson – I would not have gotten started in the first place. Basic Books made a commitment, and Elizabeth Maguire, the editorial director of Basic, and William Morrison were deft and thoughtful editors. *The New York Times*, and especially the executive editor, Joseph Lelyveld, generously gave me a leave of absence – an absence that lasted longer than advertised.

Tom Goldstein, the dean of Columbia Graduate School of Journalism, kindly offered me a place to work from while I was away from the *Times.*

I am most grateful to the Alfred P. Sloan Foundation and Doron Weber, director of the foundation's program for the public understanding of science and technology, for a grant to help me complete the book.

I want to thank the professional and educational organizations that helped me in my research. The Association for Computing Machinery gave me access to its digital library, which was invaluable. The Charles Babbage Institute's oral history project was another important resource. The videotaped lectures sponsored by the Computer Museum History Center were also extremely useful.

I am especially thankful to the people who extended their time, patience, and wisdom in interviews. These people suffered all manner of inquisition and harassment in lengthy interviews in person – sometimes repeatedly – and by phone and e-mail. Some even reviewed portions of the manuscript. They include: Alex Aiken, Fran Allen, Dennis Allison, Marc Andreessen, John Backus, Jean Bartik, Brian Behlendorf, Robert Bemer, Tim Berners-Lee, Dan Bricklin, Frederick P. Brooks, Jr., Tom Button, Martin Campbell-Kelly, Peter Capek, Steve Capps, Don Chamberlin, Alan Cooper, George Coulouris, Richard Dawkins, Doug Engelbart, Bob Frankston, John Gage, Bill Gates, Richard Goldberg, James Gosling, Jim Gray, Satish Gupta, Lois Haibt, Andy Hertzfeld, Anders Hejlsberg, Tony Hoare, and Watts Humphrey.

They also include Bill Joy, Philippe Kahn, Howard Katz, Alan Kay, Ken Kennedy, Brian Kernighan, Don Knuth, Thomas Kurtz, J. A. N. Lee, Butler Lampson, John McCarthy, Pamela McCorduck, Dan McCracken, Douglas McIlroy, Roger Needham, John H. Palmer, Raj Reddy, Dennis Ritchie, Jean Sammet, David Sayre, Eric Schmidt, Carl Shapiro, Fred Shapiro, Mike Sheridan, John Shoch, Charles Simonyi, Richard Stallman, Guy Steele, Bjarne Stroustrup, Randy Terbush, Charles Thacker, Ken Thompson, Linus Torvalds, Joseph Traub, Guy Tribble, Arthur Van Hoff, Maurice Wilkes, Irving Wladawsky-Berger, Richard Saul Wurman, and Irving Ziller.

I would also like to thank John Markoff, my friend and colleague, for his counsel, lodging in San Francisco, and his patience, because he could not depart for his book leave until I returned. Thanks also to John's wife, Leslie Terzian, for welcoming a frequent visitor.

And finally, thanks to Terry and Nikki, as always.

1

Introduction: The Rise of Software and the Programming Art

A LONE SAILBOAT IN THE DISTANCE makes its way across the rippled surface of Lake Washington in the crisp autumn dusk, framed on the horizon by the skyline of Seattle. The view is from the lakeside home of Charles Simonyi, who was a 17-year-old computer programming prodigy when he left Budapest for good in 1966. Since then, he has come a remarkable distance, in every sense. His house, though all but invisible from the road, sweeps down the hillside toward the water's edge, covers more than 20,000 square feet and includes a library, computer lab, fitness center, and swimming pool. Made of glass, wood, and steel, the home is a work of high modernism, outside and in. The black floors of polished stone glisten, and visitors are asked to remove their shoes. The walls are bare except for works of modern art by Roy Lichtenstein, Jasper Johns, and Victor Vasarely. Besides art, Simonyi collects jets. He has two, including a retired NATO fighter, which he flies. His multimillion-dollar philanthropic donations have placed his name on an endowed chair at Oxford University and on the mathematics building at the Institute for Advanced Study in Princeton. Simonyi fled Hungary as a teenager with nothing, but he now regards money with the nonchalance of the billionaire he has become. "I have no mercenary reasons for things anymore," he said.

Simonyi owes it all to software, and his uncanny facility with computer code – aided, of course, by good timing, good luck, and the whimsy of capitalism. His

1

career began at Hungary's Central Statistical Office in the mid-1960s, where he was a kind of communist version of an American teenage computer hacker. He hung around, made himself useful, and taught himself how to program on a Russian-made Ural II. In computing time, the Budapest center was living in the early 1950s, generations behind the West. Over the years, advances in software have allowed programmers to lift their gaze up further and further from the level of binary digits, or bits – the 1's and 0's that are the natural vernacular of the machine. But Simonyi learned to talk to the computer almost entirely on the machine's terms. "It was Stone Age programming," he recalled. "I've been through a time warp."

After immigrating to the United States, Simonyi changed his name from Karoly to Charles. He attended the University of California at Berkeley and Stanford University, and later joined the Xerox Palo Alto Research Center. Simonyi was at Xerox PARC during the glory years of the 1970s, when the team there did so much of the research and development that has shaped how people use personal computers. At Xerox PARC, Simonyi was the principal developer of Bravo, an innovative program for writing and editing text that allowed a person to display words on a computer screen as if plucked from the imagination of a skilled typesetter. It was a capability that became known as WYSIWYG – "What You See Is What You Get" – and it opened the door to the desktop publishing industry, and helped define the personal computer as a tool for enhancing individual creativity.

When it became clear that Xerox did not really grasp the significance of the work of its Palo Alto lab, Simonyi looked for work elsewhere. In the summer of 1980, he made an unannounced call, on a little company outside Seattle trying to make its way in the fledgling personal computer industry – Microsoft. The startup had only 40 employees, but Simonyi sniffed the future there. He and Bill Gates hit it off immediately, and Simonyi went to Microsoft.

Microsoft's Word text editor is one of the most widely used software programs in the world, and Simonyi is the "father of Word," the commercial descendant of Bravo. To him, the personal computer is a kind of delivery vehicle for software, empowering users and magnifying the power of the programmer. "You write a few lines of code and suddenly life is better for a hundred million people," he said. "That's software."

For the last several years, Simonyi has been working on an ambitious research project with the goal of greatly improving the productivity of computer programmers. He believes that the tools and methods programmers use are still fairly crude, limiting the amount of human intelligence that can be transmitted in software and thus slowing progress. Despite the constraints, Simonyi cannot help but marvel at the rise of software during his lifetime. "It shows how powerful software is. Even with the primitive tools we still use, look at how much software can do. It's amazing."

The ascent of software in the postwar years – as a field of endeavor, as an industry and as a medium of communication and commerce – has been rapid, remarkable, and almost surreptitious. The ancestry of what we now call computer programming goes back at least to the nineteenth century, when the English mathematician Charles Babbage struggled with how to handle calculations in his Analytical Engine, a conceptual forerunner of the modern computer. What he was trying to do we would now call programming. The most fundamental concept in programming is the algorithm – simply put, a set of instructions for doing something, a recipe for calculation. The algorithm apparently traces its roots to the Babylonians, and the word is a distortion of al-Khwarizmi, the family name of a Persian scholar, Muhammad ibn Musa al-Khwarizmi, who wrote a treatise on algebraic methods.

Yet it was not until World War II that electronics had advanced to the point that building useful computers became a real possibility. In those early days, programming was an afterthought. It was considered more a technician's chore, usually referred to as "setting up" or "coding" the machine. The glamour was all in the hardware – that was deemed real science and engineering. The ENIAC, for Electronic Numerical Integrator and Computer, was the machine generally credited with starting the era of digital electronic computing. That computer, at the University of Pennsylvania, did not have software. Its handlers had to set up the machine by hand, plugging and unplugging a maze of wires and properly positioning row upon row of switches. It was as if the machine had to be rebuilt for each new problem. It was hard-wired programming. To do it, the government hired a handful of young women with math skills as trainees. These early women programmers were known, literally, as "computers," a throwback to the eighteenth century use of the term to refer to the human computers who prepared statistical tables used in map-making and ocean navigation.

Programming the ENIAC to calculate the trajectory of artillery shells – its

Pentagon-assigned mission – was painstaking and difficult work, and the women devised some innovative techniques for simplifying the process. They would draw elaborate charts on paper, mapping out how the problem could most efficiently navigate its way through the machine. Then, they would set up the machine by hand. "We knew how every wire and every switch was to be set," recalled Jean Bartik. That could take weeks. Yet, thanks to their efforts, the ENIAC's public demonstration was a great success. It could calculate a firing trajectory faster than a shell flew. "Fabulous," Bartik recalled, "one of the most exciting days of my life," though it was in the spring of 1946, after the war was over.

The term used to describe the practitioners of the new profession evolved quickly. A human "computer" became a "coder." And "programmer" would soon irresistibly supplant the more quotidian label – apparently a contribution from some English members of the craft, perhaps being both more status-conscious and more literary. Grace Hopper, a software pioneer who began computing equations for the war effort on the Harvard Mark I in 1944, always felt that programming was too lofty a term for the early work. "The word 'programming' didn't appear until it came over from England," she recalled. "Actually I think what we were writing when we wrote machine code was coding. We should have reserved the word programming for a higher level. But it came over from England, and it sounded better than being a coder so everyone wanted to be a programmer."

Higher-level programming, however, would soon be possible because of a breakthrough in computer design. The idea came out of the ENIAC group, and was articulated in a June 1945 paper, "A First Draft of a Report on the EDVAC," written by John von Neumann. A renowned mathematician and game theorist, von Neumann was a consultant to the Manhattan Project that developed the atomic bomb. Designing the bomb required thousands of computations, mostly done by battalions of clerks with desktop calculating machines. So von Neumann, intrigued by the potential of computers, became a consultant to the ENIAC project in 1944. The EDVAC, for Electronic Discrete Variable Automatic Computer, was to be the successor to the ENIAC. Others were involved in the EDVAC planning, notably the ENIAC project leaders, J. Presper Eckert and John Mauchly, but von Neumann wrote the report and he got the credit for designing the "stored-program computer," which later became known as the von Neumann architecture. Virtually every computer today employs the von Neumann architecture.

The early stored-program computers began appearing after the war. The stored-program design meant that not only the computer's data – typically, numbers to be calculated in those days – but also its programming instructions could be stored in the machine. At one level, there was a straightforward efficiency benefit to this, enabling a measure of automation. The hand work of setting switches and wires could be eliminated because the programming instructions could be placed onto punched cards or tapes and fed into the computer, along with the data to be processed.

Yet there was a much deeper implication to the stored-program concept. It would make building software an engineering discipline that, in the phrase of the computer scientist Butler Lampson, is "uniquely self-referential" in that all the machinery of computing could be applied to itself. That is, a stored-program computer could be used to have programs modify other programs or to create new ones. And it is this computer-mediated interaction of programming code – a digital ecology inside the machine, one piece of code scooting off, modifying another piece, which loops back to mingle with yet another – that made possible the development of programming languages that are far more understandable to humans than binary 1's and 0's. This ability of code to assemble, reassemble, and modify itself constantly is behind everything from computer games to the Internet to artificial intelligence.

The developers of the early stored-program computers were also the first to get a real taste of the intricate, often unforeseen complexity of programming. The first stored-program computer to get up and running was built by a team led by Maurice Wilkes at Cambridge University. The machine was called the EDSAC, for Electronic Delay Storage Automatic Calculator. In his memoir, Wilkes recalled precisely when he first grasped that "bugs" were destined to be the programmer's eternal nemesis. "By June 1949," Wilkes wrote, "people had begun to realize that it was not so easy to get a program right as had at one time appeared." Wilkes was laboring to get his first "non-trivial program" to work and as he was about to mount a flight of stairs at Cambridge, he remembered, "the realization came over me with full force that a good part of the remainder of my life was going to be spent in finding errors in my own programs."

The word "software" arrived on the scene long after computers were in use, suggesting a grudging recognition of this troublesome technology. The first published use of "software" as a computing term was in 1958, in the *American*

Mathematical Monthly. John Tukey, a mathematician at Princeton University, wrote, "Today the 'software' comprising the carefully planned interpretive routines, compilers, and other aspects of automative programming are at least as important to the modern electronic calculator as its 'hardware' of tubes, transistors, wires, tapes and the like." Such sentiments were not necessarily the prevailing view at the time.

In the engineering culture of computing, programmers were long regarded askance by the hardware crowd; hardware was the real discipline, while programmers were the unruly bohemians of computing. The hardware people tended to come from the more established field of electrical engineering. There were EE departments in universities, and hardware behaved according to the no-nonsense rules of the "hard sciences" like physics and chemistry. Some mathematicians were fascinated by computers and programming, but their perspective was often from the high ground of theory, not wrestling with code and debugging programs. It was not until the 1960s, with the formation of computer science departments, that programming began to be taken seriously in academia, and then only slowly.

The recruiting and hiring of programmers in the 1950s, and beyond, was scarcely a science. Programming skills were much in demand: new people had to be trained, but there was no sure test for ability. "Early programming is where the story originated that if you looked in one ear and couldn't see daylight you could hire the person," said Robert Bemer, who was a manager in IBM's programming research department in the late 1950s. "It seemed we were just taking personnel in off the streets." Lois Haibt joined IBM in 1955, becoming a member of the 10-person team that developed the Fortran programming language as a freshly-minted graduate from Vassar College. "They took anyone who seemed to have an aptitude for problem-solving skills – bridge players, chess players, even women," she recalled. As an IBM manager, Bemer cast his recruiting net broadly. "I once decided to advertise for chess players because I thought they would be pretty good programmers. It worked very well. We even hired the US chess champion, Arthur Bisguier. He mostly played chess and didn't do that much programming." Lesser chess players, however, proved to be more productive. The ads in 1957, which appeared in *The New York Times, The Los Angeles Times,* and *Scientific American,* yielded four or five hires – a good catch, Bemer figured, at a time when there were an estimated 15,000 professional programmers in the United States, roughly 80 percent of the world's code writers.

Today, much has changed. The software industry is huge, employing nearly

9 million professional programmers worldwide. Computer science is a respected field in academia; fine minds and research funding are dedicated to plumbing the mysteries of software. For good reason, since it is software that animates not only our personal computers and the Internet, but also our telephones, credit-card networks, airline reservations systems, automobile fuel injectors, kitchen appliances, and on and on. A presidential advisory group on technology observed in 1999 that software is "the new physical infrastructure of the information age" – a critical raw material that is "fundamental to economic success, scientific and technical research, and national security."

Indeed, the modern economy is built on software, and that dependence will only grow. Business cycles and Wall Street enthusiasms will come and go, but someone will have to build all the needed software. Programmers are the artisans, craftsmen, brick layers, and architects of the Information Age. None of this could have been imagined in the early days, because no one could foresee what the pace of technological change would make possible – the ever-expanding horizons of computing, thanks to advances in hardware and software. John von Neumann and Herman Goldstine, leading computer visionaries of their day, wrote in 1946 that about 1,000 lines of programming instructions were "a reasonable upper limit for the complexity of problems now envisioned." An electric toothbrush may now have 3,000 lines of code, while personal computer programs have millions of lines of code.

Despite its importance, computer programming remains a black art to most people, and that is hardly surprising. Software, after all, is almost totally invisible. It cannot be touched, felt, heard, smelled, or tasted. But software is what makes a computer do anything useful, interesting or entertaining. Computers are very powerful, but very dumb, machines. Their view of the world is all 1's and 0's, switches ON or OFF. The simple computer that ran the "Pong" video game of the 1970s – two lines of light for "paddles" tapping a cursor-like "ball" – saw the world like this:

001110101010100001110001101010101000

And IBM's Deep Blue supercomputer, which defeated the world chess champion Gary Kasparov in 1997, saw the world like this:

001110101010100001110001101010101000

There were, fundamentally, only two differences between those two computers. The superior speed and power of the turbocharged bit-processing engine in Deep Blue, and the software. Software is the embodiment of human intelligence – the mediator between man and machine – conveying our questions or orders to the computers that surround us.

As a profession, programming is a curious blend of art, science, and engineering. The task of making software is still a remarkably painstaking, step-by-step endeavor – more handcraftmanship than machine magic, a form of creativity in the medium of software. Chefs work with food, artists with oil paint, programmers with code. Yet programming is a very practical art form, and the people who are pulled to it have the engineering fascination with how things work and the itch to build things.

As a child, Grace Hopper would tear apart and rebuild clocks. Ken Thompson, creator of the Unix operating system, built backyard rockets. Dan Bricklin, co-creator of the electronic spreadsheet, built the family television from a Heathkit set. James Gosling, creator of the Java programming language, rebuilt old farm machinery in his grandfather's yard in Calgary. Building things, it seems, is the real thrill for those naturally drawn to programming – especially so since software is a medium without the constraints of matter. The programmer can build simulated cities without needing steel, glass, or concrete; simulated airplanes without aluminum, jet engines, or tires; simulated weather without light, heat or water. At a computer, the programmer can make ideas real – at least visually real – and test them in a virtual world of his or her own creation.

Much of the history of computer programming can be seen as the effort to extend the franchise – to make it easier for more and more people to program. FORTRAN, the first real programming language, was intended to make it easier for scientists and engineers to program. COBOL was designed to make it easier for business people to program. Over the years there have been a succession of advances in programming to make things less difficult. But the idealistic vision of making programming accessible to everyone – a notion that first surfaced in the 1960s – has remained out of reach, although there have been significant strides. Nearly everyone can use a computer these days, and many thousands, even millions, of people can do the basic programming required to create a Web page or set up a financial model on a spreadsheet.

Yet more serious, and seriously useful, programming remains a fairly elite activity. By now, there has been research done on skilled programmers. It has found, yes, they share certain intellectual traits. They are the kind of people who have deep, particular interests outside work as well as professionally. An interest in science fiction, for example, will tend to be focused on one author or two. The same would be true of music, recreational pursuits, whatever. It is the kind of intellectual intensity and deep focus required in programming. In psychology, academics have looked at software programmers when studying what is called flow — a state of deep concentration, total absorption, and intellectual peak performance that is the mental equivalent of what athletes describe as being in the "zone."

Still, such study only hints at what it takes, and who has the potential, to be a gifted programmer. "Some people are three to four times better as programmers, astonishingly better than others with similar education and IQ," said Ken Kennedy, a computer science professor at Rice University. "And that is a phenomenon that is not really understood" — further evidence, it seems, that programming is as much art as science.

Donald Knuth has spent his career teaching the craft. Knuth, a professor emeritus at Stanford, helped create the field of computer science as an academic discipline. He is best-known as the author of the defining treatise on writing software, *The Art of Computer Programming*, a project he began in 1962 and that now runs to three volumes, and counting. In the book-lined, second-floor study of his home in the hills behind Stanford, Knuth observed, "There are a certain percentage of undergraduates — perhaps two percent or so — who have the mental quirks that make them good at computer programming. They are good at it, and it just flows out of them. . . . The two percent are the only ones who are really going to make these machines do amazing things. I wish it weren't so, but that is the way it has always been."

This book is about a comparative handful of those people with the requisite mental quirks to build amazing things in code. It is intended as a representative — by no means definitive — history of computer programming, told mainly through the stories of some of the remarkable people who made it happen and of the software they built.

2

FORTRAN: The Early "Turning Point"

By August 1952, IBM's sleek new computer, the Defense Calculator, was ready for a road test. A half-dozen customers had placed orders – the Los Alamos nuclear weapons laboratory, Douglas Aircraft, Lockheed Aircraft, and a few others – and they were summoned to IBM's Poughkeepsie plant to get an early glimpse of what the machine could do. Computing was in its infancy, just a step or so beyond a laboratory experiment. Interest in the electronic behemoths came mainly from the Pentagon and its private-sector relation, the emerging aerospace industry. Their interest was primarily in using the giant machines to automate the tedious process of producing scientific calculations by hand – row upon row of office workers cranking away on desktop calculators. Only gradually would it be recognized that computers were capable of being far more than big adding machines – that, when properly programmed, computers could be used as tools for exploring new frontiers of knowledge.

The impetus for the Defense Calculator came from the Korean War. The Korean conflict, begun in 1950, lent urgency to the push for new planes and weapons that would operate at higher speeds, higher temperatures, and with greater precision. Designing and producing them meant another surge in demand for engineering calculations, only five years after the end of World War II. The Pentagon and its corporate suppliers were sophisticated customers with deep pockets, but they were few. And it was not yet clear that there would be a lucrative market for the tireless calculating capacity of electronic computers outside the defense establishment.

Within IBM, there were two disparate schools of thought. The enthusiasts, led by Thomas Watson Jr., the scientists, and younger managers, understood that the demand for computing would spread broadly – and Remington Rand's UNIVAC computer was showing the way, having sold a machine to the Census Bureau. The skeptics at IBM included the chairman Thomas Watson Sr., and much of the senior management. Customers would be scarce, they worried, and manufacturing such a technically challenging machine would drain the company's engineering resources. The plan for the new machine was approved in early 1951, but in deference to the in-house skeptics the machine was called the Defense Calculator, suggesting that it was a special project in support of the war effort.

Despite the name, the Defense Calculator was a stored-program computer, and so it was a general-purpose machine, awaiting only programming instructions to tackle all manner of problems. Indeed, by the time the machine was unveiled to the public in April 1953 it had undergone a name change, becoming the IBM 701 – the first of the 700 series which firmly set IBM on its way to being the world's dominant computer maker. The 701 was compact and stylish by the computing standards of the time. The system was a collection of stand-alone units that looked something like a department-store display of 1950s-vintage kitchen appliances – the pair of tape readers resembling big cabinet-style televisions; the printer, an oven; the cathode-ray storage unit, a refrigerator. Yet it was the speed of the 701 for its day that most impressed the private audience who gathered in Poughkeepsie in the summer of 1952. They brought with them sample programs, encoded and punched onto paper tape. "They each got a shot at the computer," recalled Cuthbert Hurd, an IBM executive who was there. "They would feed a program into the computer and, bam, you got the result. . . . We all sat there and said, How are we going to keep this machine busy? It's so tremendously fast. How are we going to do that?"

The mammoth, costly IBM machines of the 1950s, to be sure, possessed a tiny, tiny fraction of the computing firepower of even a handheld computer today. But the 701 was a speed demon in 1952, so IBM found itself facing the digital paradox – the total interdependence of two very different disciplines, computer software and computer hardware, the yin and the yang of computing. The answer to Hurd's question about how to keep the fast computer busy was simple enough: put more problems on the machine. But there was a bottleneck, and it was programming.

Preparing an engineering or scientific problem so that it could be placed on a computer was an arduous and arcane task that could take weeks and required special skills. Only a small group of people had the mysterious knowledge of how to speak to the machine, as if high priests in a primitive society. Yet there were some heretics in the priesthood, and one of them was a young programmer named John Backus. Frustrated by his experience of "hand-to-hand combat with the machine," Backus was eager to speed things up and somehow simplify programming. "I figured there had to be a better way," he recalled nearly five decades later at his San Francisco home, which overlooks the Golden Gate Bridge. "You simply had to make it easier for people to program."

In late 1953, Backus sent a brief letter to Hurd, asking that he be allowed to search for a "better way" of programming. Hurd gave the nod and thus began a research project that would eventually produce, in 1957, a historic breakthrough in computer programming, a language called FORTRAN. The managerial touch was light and the working environment informal. Backus never made a formal budget, even as the project grew and the timetable for completion slipped again and again. The team that created FORTRAN would build gradually, one by one, until it reached 10 people. It was a young group, all still in their twenties or early thirties when FORTRAN was released. The team was heavy with math training because so much of computing at the time was numerical analysis and mathematics, sorting through all those numbers.

Still, it was an eclectic bunch – a crystallographer, a cryptographer, a chess wizard, an employee loaned from United Aircraft, a researcher from MIT, a young woman who joined the project straight out of Vassar. They worked together in one open room, their desks side by side. They often worked at night because it was the only time they could get valuable time on the machine to test and debug their code. The odd hours and close work bred camaraderie. For relaxation, there were lunch-time chess matches and, in the winter, impromptu snow ball fights. They knew each other, and they knew their code intimately and the machine they were working on, right down to the metal. And they were outsiders to the industry establishment, which regarded their chances of success as slim to nil. "We were the hackers of those days," Richard Goldberg recalled at the age of 76.

The success of the FORTRAN team was twofold. First, they devised a programming language that resembled a combination of English shorthand and algebra. It was a computing vernacular that was very similar to algebraic formulas that scientists and engineers used daily in their work. So FORTRAN opened up programming to the people whose problems were being put on computers in those days. With some training, they were no longer dependent on the computing priesthood to translate their problems into the language of the machine. FORTRAN moved communication with the computer *up* a level, closer to the human and away from the machine. That is why FORTRAN is called the first *higher-level* language.

But the greater achievement of FORTRAN was that it worked so well. That is, FORTRAN generated programs that ran as efficiently, or very nearly as efficiently, as ones hand-coded so painstakingly by the programming elite. Without that leap in programming automation, FORTRAN would have never been adopted. Machine time was a precious, costly resource. If programs written in FORTRAN had run slowly, consuming far more machine time than hand-coded programs, it would have been economically impractical. Matching the run-time efficiency of human programmers was thought to be impossible at the time. Yet the IBM team succeeded because of their masterful design of the FORTRAN compiler. Put simply, a compiler is a program that captures the human intent of a program and recasts it in a way that is understandable – executable, that is – by the machine.

Modern versions of the FORTRAN language are still widely used for some scientific computing tasks – for the numerical analysis work involved in weather prediction, modeling changes in the climate, and in high-energy physics, for example. Yet today, FORTRAN is often mentioned by experienced computer scientists and veteran programmers wistfully, as the first programming language they learned but then abandoned as newer languages developed for new kinds of computing. FORTRAN was something you grew out of. But to point out how quickly programming has moved to generations of new tools in no way lessens the extraordinary advance that FORTRAN gave to the world of software. Other programming languages rose from the foundation that FORTRAN built. J. A. N. Lee, a professor at Virginia Tech and the dean of computer historians, has called FORTRAN "the turning point" in the development of programming languages and its compiler technology – the software equivalent of the transistor. Ken Thompson, who created the Unix operating system at Bell Labs in 1969, observed that "ninety-five percent of the people who programmed in the early years would never have done it without FORTRAN. It was a massive step." Or, as Jim Gray, a leading software researcher who now works for Microsoft, declared with a certain biblical flourish, "In the beginning, there was FORTRAN."

John Backus had followed a haphazard path to computer science. He was raised in Wilmington, Delaware, the son of a self-made man, Cecil Backus, who was trained as a chemist but switched careers to become a stockbroker. The elder Backus prospered as a partner in a brokerage house, and the family became wealthy and socially prominent. As a child, Backus enjoyed experimenting with his beloved chemistry set. He recalled with satisfaction the time when he was about 12 when he revived a motorbike that another youngster had given up for dead after it careened into the ocean. "I've always liked mechanical stuff," observed Backus, a rail-thin man with close-cropped gray hair and a self-deprecating manner.

At 76, Backus cheerfully described himself as still "a gadget freak." He confesses an addiction to his Palm Pilot. "Couldn't live without it," he joked. He has rigged up his own automatic remote controls for his front gate and garage door. He had just acquired a television set-top gadget that, using a large computer disk and some clever programming, allows viewers to skip commercials, pause while viewing live broadcasts and record television programs based on database searches. "This is a great invention," Backus declared with delight. "It's going to change television."

Backus had a complicated, difficult relationship with his family, and was a wayward student. His parents sent him to an exclusive private high school, The Hill School in Pottstown, Pennsylvania. His grades were so poor that he was sent every summer to a study camp to allow him to advance with his class the following fall. "I loved the fact that flunking courses meant I did not have to go home," Backus said. He regarded The Hill School as a problem-solving challenge of sorts. "The delight of that place was all the rules you could break," he said. His approach to formal education was unchanged in college. He lasted two semesters at the University of Virginia before he flunked out.

His uninspired performance as a student had nothing to do with his intellect, as his military experience soon demonstrated. Backus was drafted in 1943, immediately after he and the University of Virginia parted ways. Stellar scores on Army aptitude tests resulted in Backus being sent first to the University of Pittsburgh for an engineering course, and later to Haverford College for a pre-med course. Next, Backus attended New York Medical College in Manhattan on a government-funded program, though he found medicine boring. "It seemed to be all memorizing procedures and body parts," he recalled.

While wondering what to do next, Backus, a classical music enthusiast, decided that what his small Manhattan apartment needed was a good sound system. He began to construct his own high-fidelity set, and soon found himself attending courses at a school for radio technicians. To build an amplifier, Backus had to calculate points on the curve of sound waves. He found wrestling with the mathematics to be difficult but compelling. "It was so awful to do that calculation, but somehow it kind of got me interested in math," he said. So Backus applied to Columbia University, which admitted him as a probationary student, given his decidedly mixed record as a scholar. He did well at Columbia, completing a bachelor's degree and earning a masters in mathematics in 1950.

One spring day, shortly before graduating from Columbia, Backus visited IBM's headquarters on 57th Street and Madison Avenue. He had heard about the massive scientific calculating machine on display there and, given his fascination with mechanical things, he wanted to take a look at it. IBM had installed the computer on the ground floor, so it could be seen from the street as a kind of tourist curiosity. With its thousands of flashing lights, clacking switches, punched cards shuffling and paper tapes whirring, the computer struck many passersby an electronic Rube Goldberg contraption. The passing pedestrian throngs probably did not know what to make of the machine, but they displayed an impulse – one repeated again and again over the years – for giving computers anthropomorphic nicknames. They dubbed the machine "Poppa."

Eager for a closer look, Backus ventured inside and he was given a brief tour and an explanation of the various parts of the Selective Sequence Electronic Calculator, known by its acronym SSEC. He mentioned to the woman showing him the machine that he was looking for a job, and that he was a graduate student in mathematics at Columbia. At that, she said she would take him straight up to see Robert (Rex) Seeber, co-inventor of the SSEC. Backus protested. "I wasn't wearing a tie, I had a hole in the sleeve of my jacket, and I didn't know anything really about computers," he recalled. No, no, not a problem, the woman insisted, and she ushered him up to see Seeber. After a brief greeting, Seeber proceeded to ask a series of questions that Backus described as "brain teasers" – such as how to handle the alignments and additions when using a 10-digit calculator to multiply a 20-digit number.

Backus recalled it as an informal oral examination, with no recorded score. Seeber hired him on the spot. As what? "As a programmer," he replied, shrugging. "That was the way it was done in those days."

Backus joined IBM during a period of rapid transition for the company, the industry, and the technology of computing. For decades, manufacturers like IBM, Remington Rand, Burroughs, and NCR had thrived mostly by producing accounting machines for business. These calculators helped managers track payrolls, inventories, and sales as large companies proliferated – enterprises that evolved in response to the economies of scale made possible by the rise of mass production, modern rail and auto transportation, and the growth of a national telephone system. But World War II had pushed the technology and the calculator makers beyond electromechanical machines for business and toward high-speed electronics for the aerospace and defense markets.

The SSEC, which IBM called the Super Calculator, reflected those trends. It was essentially an IBM science project, a one-of-a-kind machine designed and built to let IBM's researchers push the limits of electronic calculators and gain experience.

The SSEC was not a stored-program computer, but it was the state of the art when it was completed in 1948, probably the most powerful computational machine at the time. The very name "computer," however, was regarded with concern by Thomas Watson Sr. The term was still often used to refer to human clerks doing calculations and, Watson worried, might fan popular fears that the new technology would mean lost jobs. The elder Watson's reluctance to use the term "computer" was understandable. Public anxiety about computers causing unemployment continued for years. In 1957, a popular movie "Desk Set," starring Spencer Tracy and Katherine Hepburn, tapped that nerve with a romantic comedy about the arrival of a computer in a big company. Bunny Watson (Hepburn) is convinced that workers in her department will be replaced by the EMERAC computer and that Richard Sumner (Tracy) is there not only as Emmy's minder, but also to slash the work force. Naturally, Bunny's worries prove unfounded, and the movie ends as movies always did in the 1950s.

Backus got his introduction to programming on the SSEC. The programs he worked on were large scientific calculations, like a program to calculate the position of the moon and nearby planets at any time over years, which required endless crunching of coefficients. "It was pure science," Backus recalled.

The research may have been lofty, but the programming was the equivalent of trench warfare. For like the wartime ENIAC, the SSEC had to be reconfigured for each assignment it was given. After figuring how to set up the problem mathematically, the researchers then had to put it onto the machine. This involved devising elaborate flow charts of how the calculations should be funneled through the machine in general terms. Next came the arduous chore of mapping out the calculating steps, instruction by instruction, intricately on preprinted sheets of paper. Then, the machine had to be set up by hand for each batch of calculations – which switches to flip and which wires to plug into which circuits, to get the Super Calculator flickering and clacking again.

On the SSEC, a large program could take months to map out, and then run on the machine for six months. It would grind to a halt every three minutes on average, requiring further ministrations from the programmers. "As a programmer, you had to be there the whole time," Backus said. When problems surfaced, the clues were to be deciphered by reading the binary coded numbers off the machine's thousands of lights.

Debugging the machine was also done by ear. Circuits were opened and closed by relays – metal bars attached to springs that were raised by the pulling force of electromagnets. The thousands of relays being slapped into position in various sequences made a deafening racket at times, yet it was not merely random industrial noise. To the trained ear of a programmer, the repeated rhythm from one corner of the machine, signifying a program was frozen in some calculating loop, was as dissonant as listening to a broken record. Later, when the next-generation 701 Defense Calculator arrived, with its mute electronic switches instead of mechanical relays, Backus recalled feeling a twinge of panic. "I wondered, 'How are we going to debug this enormous silent monster.' "

Backus spoke of his days wrestling the Super Calculator with a sense of nostalgia for the frontier. "Oh, the machine was so complex. It was. And there was no textbook back then. The constraints were such a challenge. . . . There was so much opportunity for ingenuity. You were inventing all the time."

In the 1953 letter to his boss proposing his programming project, Backus emphasized the economic dimensions of the problem. At most installations, the cost of programmers' salaries in computer centers – typically 30 programmers per installation – was at least equal to the cost of the computer (the monthly rental for the 701 was $15,000, the equivalent of nearly $100,000 these days). In addition, Backus noted, one-quarter to one-half of computing time was spent in debugging. Accordingly, programming and debugging represented as much as three-quarters of the cost of operating a computer. With hardware improving rapidly and becoming cheaper, the proportionate cost of programming seemed destined to rise even further. It was a big problem, and getting more so. Cuthbert Hurd read the letter and immediately approved the request to begin the programming research project. "He really understood," Backus recalled.

In January 1954, Backus got his first conscript, Irving Ziller. A graduate of Brooklyn College, Ziller joined IBM in 1952 and had been put to work programming "plug boards" on electronic calculators. The calculators were made from a series of these plug boards, roughly 8½ by 11 inches, filled with holes into which wires were connected by hand. It was another form of hard-wired programming. When complete, a plug-board would look like a miniature jungle of wires rising up from the board. Ziller quickly proved to be both bright and extremely adept as a plug board programmer. In his apartment in the Riverdale section of New York, Ziller described his plug-board programming days in animated detail. "This, as you can imagine, was a fairly tedious job," he said. "Anyone doing plug boards understood the emerging need to simplify the programming process." So, when asked, Ziller was an eager recruit to Backus' project.

Soon after, the team got its third member, Harlan Herrick. He was a math major at Iowa State University, and an outstanding chess player, who had won regional tournaments in the Midwest. He was awarded a scholarship to Yale University for graduate studies, but he was unhappy there. After reading an article about IBM's SSEC machine, he applied for a programming job and was hired.

When he joined the FORTRAN team, Herrick had five years of experience programming IBM's SSEC and 701 machines. That made him a wizened veteran among programmers at the time. Within IBM, Herrick was known as a naturally gifted programmer, and his work was instrumental to the success of FORTRAN. At the start, though, he was the most skeptical because he was the most steeped in the programming practices of the time. Herrick was a member of the priesthood. When Backus first told him about the project, Herrick was incredulous. "I said, 'John, we can't possibly simulate a human programmer with a language – this language – that would produce machine code that would even approach the efficiency of a human programmer like me, for example,'" Herrick recalled in 1982. "I'm a great programmer, don't you know?"

Given the intellectual rigors of their craft, programmers of the day were understandably disbelieving, even disdainful, of a programming language doing their jobs. They had to be conversant in the machine's tongue, in binary. For a flavor of the simplest numeric translation, 1 is 01 in binary, 2 is 10, and 3 is 11. Then 4 is 100, because it is 2 squared, requiring that a digit be added in the third column. The columns, moving to the left, are "powers," or multiples, of 2 – so 1000 is 8, which is 2 cubed ($2 \times 2 \times 2$). And 256 is 100000000, or 2 to the eighth power. The binary system of 1's and 0's is, at first, perplexing to humans, somehow "unnatural." But, in part, that is because we are so accustomed to the number system based on 10, with the number columns being powers of 10. The base-10 system – called decimal – also feels comfortable because it corresponds to the natural human counting tool of our 10 fingers – our digits.

An early tool to simplify things for programmers staring at a blizzard of binary was the use of octal notation. Octal is a base-8 number system, which uses eight symbols (0, 1, 2, 3, 4, 5, 6 and 7), and its columns moving left were powers of eight. Octal was used because it was relatively easy for humans to read – certainly easier than binary – and yet could be easily translated into the binary format of the machine because, again, eight is a power of two. For early programmers, octal became second nature. "We used to joke that we did our checkbooks in octal," said Lois Haibt, a member of the FORTRAN team. There was even octal humor. "Why can't programmers tell the difference between Christmas and Halloween? Because 25 in decimal is 31 in octal." The joke's answer, when written, became: Dec(imal) 25 = Oct(al) 31. (That is, 31 in octal, or 3×8 plus 1, is 25 in decimal.) Since the 1960s, as computers and software became larger and more complicated, programmers have typically used a base-16 system, called hexadecimal, as a shorthand for binary when they really have to understand things at the machine level. In hexadecimal, the symbols used are 0 through 9 and A through F.

The next step in trying to make the programming process less arduous was the development of "assembler" programs. These allowed programmers to write instructions using mnemonic abbreviations – perhaps LD for "load" or MPY for "multiply," followed by a number to designate a location in the computer's memory. A small assembler program then translated, or "assembled," these symbolic programming instructions into binary so the machine could execute those instructions. The symbolic shorthand – the blend of abbreviations and numbers – was called an assembly language. Each different kind of computer had its own assembly language, as if each machine environment were a medieval fiefdom with its own dialect. Still, the assembly languages with their assembler programs were an essential step on the way toward higher-level languages like FORTRAN and its compiler.

The assembler was pioneered in England, where the first working stored-program computer went into operation, Cambridge University's EDSAC. The programming innovation in Cambridge was inspired by the same thinking that would motivate the FORTRAN team and generations of software developers afterwards. "The objective from the very early days was to make it easy to use for people without specialized training," recalled David Wheeler, who was a 21-year-old researcher when he joined the Cambridge group in the fall of 1948. Wheeler wrote the assembler program for the EDSAC, which he called "Initial Orders," an artful and elegant 30 lines of instructions. The Initial Orders program would translate into binary the instructions written in a simple assembly language. A single line of instruction to tell the computer to "add the number in memory location 123 into the accumulator" would appear:

A   123   F

The Cambridge group described their work in the first programming textbook, *The Preparation of Programs for an Electronic Digital Computer,* published in 1951. The authors – Maurice Wilkes, David Wheeler, and Stanley Gill – chose to have the book published first in the United States, where there was a larger computing community and their work might have the most impact. The book also described the use of "subroutines" – segments of programs that are frequently used, so they can be kept in "libraries" and reused as needed in many software applications. The Cambridge group thus introduced the concept of reusable code, which remains today one of the principal tools for reducing software bugs and improving the productivity of programmers.

At the insistence of Backus, the FORTRAN team was aiming far higher than the level of an assembly language. Each line of assembly code translated into one instruction of binary machine code. For the assembly programmers of the 1950s, programming was a one-line-at-a-time craft. Backus wanted to break through the one-to-one arithmetic of programming so that one line written by a human might translate into many machine instructions. His plan, if successful, would bring not just a technological advance to computing, but a certain cultural shift as well. His goal, after all, was to automate that fine, hand-crafted art of the assembly programmer.

Others were pursuing the same goal, hoping to make computing less dependent on the programming priesthood. Perhaps the most outspoken advocate for change was Grace Hopper, who worked during the 1950s for Remington Rand on the UNIVAC, for Universal Automatic Computer. She defined the "programming problem" in much the same terms as Backus. "I felt that sooner or later our attitude should not be that people should have to learn how to code for the computer," she explained in 1976. Instead, Hopper said, the computer should "learn how to respond to people because I figured we weren't going to teach the whole population of the United States how to write computer code. There had to be an interface built that would accept things that were people-oriented and then use the computer to translate to machine code."

Hopper spoke frequently at computer gatherings to marshal support for what she called "automatic programming." Under that banner, she grouped several software tools. She wrote an automatic programming system for the UNIVAC that stitched together pieces of code into a single program, and she called it the A-O compiler, and versions A-1 and A-2 would follow. But Hopper, who would later be a leader on the committee that oversaw the creation of COBOL, was always more a technologist, visionary, and industry stateswoman than a programming wizard. Her compiler produced programs that ran far too slowly for most commercial uses. And it was a collection of programming aids rather than software that meets the modern definition of a compiler: a program that translates instructions written in a language familiar to human beings into binary.

The first true compiler in the contemporary sense was probably built for the government-funded Whirlwind project at the Massachusetts Institute of Technology. A pair of MIT researchers, J. Halcombe Laning and Neal Zierler, wrote a program that translated algebraic equations into machine code in early 1954. Backus and Ziller visited MIT in June of that year to observe the compiler firsthand and speak with its creators. "It was pretty good, very nicely done conceptually," Ziller recalled. "But they took an academic approach. They couldn't care less about efficiency." In fact, programs using the MIT compiler took nearly 10 times longer to run than it took for hand-coded programs doing the same calculations.

Undaunted, the original FORTRAN trio – Backus, Ziller, and Herrick – set the ambitious goal for themselves of matching the work of human coders. Despite setbacks, they never wavered. Success or failure, they understood, would hinge on the efficiency of compiler translation far more so than the language itself. "We simply made up the language as we went along," Backus explained. "We did not regard language design as a difficult problem, merely a simple prelude to the real problem: designing a compiler that could produce efficient programs."

The approach may have been nonchalant, but the FORTRAN language design certainly left a legacy – especially the seemingly innocuous decision to include the Go To statement as one of its basic commands. In 1968, Edsger Dijkstra, an academic champion for the concept of more disciplined "structured programming," wrote an impassioned letter to the editor of *Communications of the ACM*, the journal of the leading professional society in the field, the Association for Computing Machinery. Under the playfully incendiary headline, "Go To Statement Considered Harmful," Dijkstra observed that the "quality of programmers is a decreasing function of the density of Go To statements in the programs they produce." The Go To statement, Dijkstra wrote, had had "such disastrous effects" that he was "convinced that the Go To statement should be abolished from all 'higher level' languages." By the late 1960s, when Dijkstra wrote, software programs had grown immensely in size and complexity. The Go To command is an "unconditional jump," allowing a hop from one place in a software program to anywhere else, altering the flow and control of a program's execution. In a big, complicated software program of the late 1960s, the Go To looked to Dijkstra as an invitation to disaster, allowing a programmer to write programs that hopped all over the place and make "a mess of one's code." Yet that problem was not the big one the FORTRAN team faced in the mid-1950s. The headaches of the 1960s would be created partly by the success of the Backus team in overcoming the big challenge of the 1950s.

The very name FORTRAN was testimony to the group's obsession with the translator, or compiler – an abbreviation for FORmula TRANslating system, which was later trimmed to FORmula TRANslator, as if to suggest something more substantial, not some amorphous "system." Backus came up with the name in 1954, to no great enthusiasm from his colleagues. "F-O-O-O-R T-R-R-A-A-N," said Herrick, slowly mouthing the syllables as if distasteful for an IBM-sponsored documentary film in 1982. "It sounds like something spelled backwards." Yet FORTRAN accurately described the project, and nobody could come up with a better idea, so the name stuck. So, of course, did the language, in use even today on machines ranging from supercomputers to PCs – something never imagined by Backus and the rest of the FORTRAN team. The initial goal for FORTRAN was to make programming easier on one machine – the successor to the 701 Defense Calculator, the IBM 704.

On 10 November 1954, the FORTRAN group produced a paper that described the FORTRAN language and its goals, "Preliminary Report: Specifications for the IBM Mathematical FORmula TRANslating System." Irving Ziller had held onto a copy of the original document, which he retrieved from a cardboard box in his basement. The 29-page report presented the math-laden vernacular of the language and its traffic-cop rules for handling operations – Fortran's DO, IF, GOTO and STOP commands. Despite its dry title, the report was also a shrewd, at times impassioned, marketing document. It was, in the parlance of modern business, a "vision statement," detailing Backus' plans and hopes for FORTRAN and his optimistic prediction of its impact. The report reads today as a fascinating mixture of foresight and naïveté. It reiterated the economic arguments about the high and increasing costs of hand-coded programming that Backus made to Hurd in 1953, which got the FORTRAN project started. Then, really warming to the subject, one remarkable paragraph began, "Since FORTRAN will virtually eliminate coding and debugging, . . ." The "coding" referred to was the machine code that FORTRAN compiler would generate automatically, so that was more descriptive than predictive. But eliminate debugging? The blithe prediction strikes the modern reader as quaintly amusing, given that debugging remains the bane of the programmer's existence.

In retrospect, the most powerful economic argument in the 1954 preliminary report was its theme of empowerment. FORTRAN, the report predicted, would not only increase the efficiency of programming, but also increase the pool of people who could program – a goal of software visionaries ever since.

"The amount of knowledge," the report declared, "necessary to utilize the 704 effectively by means of FORTRAN is far less than the knowledge required to make effective use of the 704 by direct coding. . . . In fact, a great deal of the information that the programmer needs to know about the FORTRAN system is already embodied in his knowledge of mathematics. Thus it will be possible to make the full capabilities of the 704 available to a much wider range of people than would otherwise be possible without expensive and time-consuming training programs."

With their marketing-and-vision document in hand, Backus, Ziller, and Herrick traveled the country, speaking to customers who had ordered a 704. They went to Washington, Los Angeles, Pittsburgh, Albuquerque, and elsewhere, espousing their vision of FORTRAN and the future of programming. They had hoped to stir up interest in the project and solicit useful suggestions from informed users, but their enthusiasm fell on deaf ears. The FORTRAN report struck the customers as wishful thinking, especially coming on the heels of the false hopes raised by "automatic programming," touted by Grace Hopper and others. Knowledgeable computing customers simply did not believe that FORTRAN – the language and its compiler – could produce machine code that approached the efficiency of hand-coded programs. Recalling those dispiriting trips, Backus observed, "We received almost no suggestions or feedback." FORTRAN, then, had little to do with the modern business truisms about the necessity of being "customer-driven" and getting users involved in product development. FORTRAN was considered too far over the horizon for customers to take seriously.

Though disappointing, the customer tour had the unintended consequence of motivating the FORTRAN team. "We thought it was a good project, and then everyone told us it couldn't be done," Backus recalled. "There was a sense that we really wanted to show them." Vindication would come, but not quickly. After the preliminary report, Backus recalled, his superiors would periodically ask when FORTRAN would be completed. He always had an answer, the same one. "Six months, come back in six months, I'd say. We honestly always felt that we were going to be done in six months from now. But it became nearly three years."

In 1955, Backus began adding to the team, and he plucked recruits from various sources. Backus visited MIT, a major center of computing research, which had a close relationship with IBM. He explained the FORTRAN project and asked if the university's Digital Computer Laboratory might want to send someone to work on FORTRAN. MIT dispatched one of its star programmers, Sheldon Best. Backus was able to exploit the one flicker of interest he got from the industry, from United Aircraft, which lent Roy Nutt to the FORTRAN project. And Nutt was a real catch – an extraordinary programmer who could "execute" a program in his head, as a machine would, and then write error-free code with remarkable frequency. Nutt, Backus marveled, would walk straight to the keypunch when he felt he had solved some software riddle and, without notes, write a flawless program onto the punched cards. Inside IBM, Backus tapped people he spotted or who were recommended. He wanted people who were bright and seemed to have knack for programming. "There was nothing formal about it," he recalled. "We added one person at a time, and it just sort of happened."

The experience of the recruits was often slender. Robert Nelson, a former cryptographer for the State Department in Vienna, was a new employee hired by IBM to do the routine work of typing scientific documents. But Backus soon recognized Nelson's technical talent. "He quickly became an outstanding programmer, absolutely crucial to the FORTRAN project," Ziller observed.

Richard Goldberg had a Ph.D. in math from New York University, and he had intended to teach. A semester at Dartmouth College convinced Goldberg he was not meant for teaching, so he moved back to New York and got a job at IBM. "I didn't know anything about computing," he recalled. But Goldberg excelled in a three-month programming course, and he was sent along to Backus. Lois Haibt went to IBM straight from Vassar College, where she recalled being "good in math and science and terrible in the fuzzy subjects like English." A scholarship student, she was more than "good" in math and science and her summer jobs were at Bell Labs. When she graduated, however,

IBM lured her with a starting salary nearly twice the Bell Labs offer – $5,100 a year, which at the time seemed a lavish sum to her. "They told me it was a job programming computers," Haibt said. "I had only a vague idea what that was. But I figured it must be something interesting and challenging, if they were going to pay me all that money." In her programming class, she proved extremely adept and was assigned to the FORTRAN team.

For David Sayre, computing was at first merely a tool which later became a career. In the early 1950s, Sayre was a crystallographer doing biophysics research at the University of Pennsylvania. His research focused on the structure of hydrocarbon carcinogenic molecules. Even then, such study of crystal structures relied on computing assistance. Frustrated by the university computer, Sayre went looking for a faster machine. He had his eye on an IBM 701. He was told time on the machine cost $400 an hour, but that for his scientific research he could get a couple of hours free. The IBM officials had markets in mind, seeing work like Sayre's as a way to make deep inroads into scientific computing. Sayre went to New York City, wrote his program line by line in octal notation and put it on the 701. He found the process intellectually satisfying and oddly compelling. "You entered a world that kind of ran the way it was supposed to, a world made for working out the logic of something," Sayre recalled. "When you ran your program the expected thing happened, and if it didn't there was a logical reason why it didn't." Sayre joined IBM, working on scientific applications but also going deeper into pure programming. He wrote a diagnostic program to help find bugs on the IBM 704, for example. So Backus "borrowed" Sayre for the FORTRAN project, and Backus had a way of holding onto the people he borrowed.

In computer science, a big part of the challenge is capturing a human problem in a way that can be worked on by the machine. The way Backus chose to break up the compiler problem, though subtle, was one of the genuine achievements of FORTRAN – and, like so many innovations, it is obvious only in hindsight. Look into most language compilers today, and one sees the same steps or phases that were in the FORTRAN compiler of the 1950s – the same problem-tackling structure Backus adopted.

The FORTRAN compiler – and the work by the team – was divided by operational tasks. The compiler first performed an initial scan, or parsing, of the higher-level language, the algebraic symbols, and English abbreviations.

Next, it performed a complex analysis of the program so that the compiler focused most of its energies on automating the working heart of the program – that is, the most frequently repeated operations in the program. Then, the compiler had to figure out how to allocate its compiling instructions to run on the machine in a way that used a minimum of computer time. And, in the last step, the compiled program had to be "assembled" into machine code.

The FORTRAN compiler accomplished all this not with brute force, but with an elegant efficiency that seemed to lend the software a certain life of its own, if not intelligence. By early 1957, when the FORTRAN project was in the home stretch and the team got compiled programs back from the computer, they were often amazed. The compiler had made what Backus called "surprising transformations" in the programs being compiled, altering the programming expressions and the order of computation. As they combed through the changes the compiler had made, they could see that the compiler's work was efficient. But, Backus said, the compiler took steps that "we would not have thought to make as programmers ourselves." In fact, the FORTRAN team was mainly observing the work of a well-designed, complex piece of software, following general rules and specific instructions embodied in the programmers' algorithms to accomplish its appointed task. Still, seeing the FORTRAN compiler's handiwork struck members of the team as a revelation that software was a special medium. "It was just amazing, the interaction of the programming instructions, almost as if the compiler was a living thing," Ziller recalled.

It is often said that the best software designers and the best programmers have an uncommon capacity for two different kinds of reasoning – conceptual and procedural thinking, high-level and low-level work. FORTRAN had both, in large doses. The language itself was a high-level, conceptual triumph. It not only made it far easier for people to program, but ensured FORTRAN's longevity. Though initially intended only for the IBM 704, FORTRAN was wisely designed, at a high enough level, so that it divorced itself from a specific machine environment. "The underlying machine could change to its heart's content and the programming language could sail along for 50 years," observed David Sayre.

FORTRAN also convincingly broke through the one-to-one arithmetic of assembly programming. A line written in FORTRAN would translate into several machine instructions, again, simplifying the programming craft. By way of example, below is a rudimentary FORTRAN program for converting temperatures in Fahrenheit to Celsius:

```
WRITE (*,*) "Please enter Fahrenheit temperature:"
READ (*,*) FAREN
CELSIUS = (FAREN − 32) / 1.8
WRITE (*,*) "The Celsius equivalent is: ", CELSIUS
STOP
END
```

In an assembly language, the same simple program runs to more than 60 lines of code. The single FORTRAN line with the conversion formula (CELSIUS = (FAREN - 32) / 1.8) becomes five lines of assembly language instructions below, in an assembler for a personal computer:

```
fld 32real [0001BEC8]
fchs
fadd 32real [0001E000]
fdiv 32real [0001BEC0]
fstp 32real [ebp-08]
```

Then, in the binary code understandable to the machine, the single FORTRAN line becomes five lines that look like this:

```
11011001000001011100100010111110000000001000000000
110010011110000
1101100000000010100000000111000000000000100000000
1101100000110101110000001011110000000001000000000
11011001010111011111000
```

The lower-level labor for the team was the long slog required to make the FORTRAN-compiled programs match the efficiency of human programmers – the goal so many in the industry thought was out of reach. It was difficult, often frustrating work. Years later, when asked about the broad lesson of the FORTRAN experience, Backus articulated a theory of innovation by iteration, a constant process of trial and error. "You need the willingness to fail all the time," he explained. "You have to generate many ideas and then you have to work very hard only to discover that they don't work. And you keep doing that over and over until you find one that does work." Their willingness to persevere in spite of setbacks and doubts owed a lot to the chemistry of the FORTRAN team. They were a young, bright, close-knit group brimming with energy and optimism. They saw themselves somewhat as outsiders to the rest of IBM, trying to do something that had never been done before in a brand-new field with few, if any, established rules. "IBM was really loose for a while in this new part of the business," Sayre recalled. "We were like a Silicon Valley operation."

The FORTRAN team was at first called the "Programming Research Group," and it was ensconced on the 19th floor of an annex to the IBM head-quarters building in New York on 590 Madison Avenue. Their office was next to the engine room for the building's elevator machinery, and their conversations were sometimes interrupted by loud mechanical rumblings next door. Indeed, their offices would remain modest and makeshift over the three-year course of the FORTRAN project, even as the team grew and then moved to the fifth floor of an office building on 56th Street. They were always a bit isolated and separate. Robert Bemer worked on the other side of the big room occupied by the FOR-TRAN group, and he mostly recalled their work regimen. "They were buried in it," Bemer said, "day and night." Irving Ziller recalled a stretch when Sheldon Best was puzzling over a particularly intractable problem. To talk it over, Ziller made a habit of walking with Best to a nearby subway station after work. Frequently, they would keep talking and walk around the block several times, before Ziller finally descended into the subway and Best strolled off to his apartment nearby.

Still, it was not all work. In the winter, snowball fights might break out in the office, with ammunition scooped off the window ledges. They would take a break once a day for coffee, donuts, and conversation at a diner around the corner. And there were the lunch time games of Kriegspiel ("War game," in German). Kriegspiel is a form of "blind chess," in which two players sit side by side, each with a board, and a divider blocks the view of each other's board. Each player makes moves in turn, and tries to imagine the moves the opponent makes. There is a referee who provides "clues," by announcing when a piece is successfully captured or when a player cannot make a move because an opponent's piece blocks the way. For a certain kind of mind, Kriegspiel was recess.

Backus organized the work in a cellular structure. Each of the groups of one, two or three people was an autonomous unit, free to use whatever techniques it deemed best for its assigned job. Yet each group also had to agree on programming specifications with its neighboring sections, so that the software code would work together smoothly. The section groups had separate tasks, but cooperation and collaboration were the norm. Lois Haibt did the "flow" analysis – essentially, predicting where the high-traffic areas of compiler would be, using a mathematical technique called Monte Carlo simulations. "It was the kind of atmosphere where if you couldn't see what was wrong with your program, you would just turn to the next person," she recalled. "No one was worried about seeming stupid or possessive of his or her code. We were all just learning together."

Ziller and Nelson had one of most difficult assignments: analyzing and optimizing the so-called inner loops of the compiler. Fine-tuning the inner loops – essential repeated operations – would be key to making the compiler efficient. Ziller and Nelson would have to automate, in software, the work that human programmers regarded as the height of their craft: squeezing instructions out of those loops.

Theirs was the procedural artistry of programming. They studied the inner loops to find the most efficient method of execution – that is, using the least machine time. They then had to devise the programming statement that would invoke that efficient step – the particular – when the compiler was presented with similar problems – the general case. "It was a repetitive process we developed, a constant iteration of improvements," Ziller explained. "We were constantly trying to save one 'load' or 'store' instruction, and juggling the order of execution so that step by step we removed more and more calculations."

Because machine time was a scarce resource, FORTRAN was debugged mainly at night. In the FORTRAN compiler, the Backus team was working on path-breaking software technology. But they used the traditional programming medium of paper tablets and punched cards, which would die only later, starting in the 1960s – thanks to hardware improvements in memory and storage technology – permitting a programmer to code on a keyboard terminal connected to the computer, and still later, directly on a personal computer. The FORTRAN programmers first wrote their code out on sheets of lined or grid-patterned paper. The programs were then keypunched onto punched cards, and the FORTRAN team did their own keypunching. Next, the cards were placed into a card-reading machine, which deciphered the digital perforations on each card. The card reader then fed the data and programming instructions into the computer, and the program ran. (FORTRAN was actually distributed to customers in late 1957 on magnetic tapes, according to David Sayre, since duplicating the big deck of cards accurately proved very difficult.) For nearly a year, the team rented rooms at the Langdon Hotel, which passed from the New York scene long ago, sleeping a little during the day and staying up all night to get time on the 704 computer in the headquarters annex nearby. At dawn each day, they retreated and placed the most resistant bugs into a folder for special attention and further work. The file was jokingly labeled "Necropolis" – a place of horror and death.

The Fortran project was managed with a light corporate touch. Still, the group did have to conform to certain IBM rules. In administering these Backus, it seems, was all but subversive. At the time, IBM conducted yearly employee reviews called the "Performance Improvement Program," or PIP, for short. The PIP, like most such programs today, followed a rigid formula, with numbers and rankings. Backus decided the PIP system was ill-suited for measuring the performance of his programmers, so his approach was to mostly ignore it. One afternoon, for example, he called Lois Haibt over for a chat. He talked about her work, said she had been doing an excellent job, and then pushed a small piece of paper across the desk saying, "This is your new salary" – a pleasing raise, as Haibt recalled. As she got up to leave, Backus mentioned in passing, "In case anyone should ask, This was your PIP."

FORTRAN was presented to the computing world in February 1957 at the Western Joint Computer Conference in Los Angeles. The gathering was mainly attended by the few dozen members of SHARE, an aptly named group of companies that used IBM computers. SHARE was made up of managers and scientists mostly from aircraft makers, big manufacturers, and government labs, and the members freely shared information, complaints, and even programs. This was before software was a separate business, so programming expertise was regarded more as useful knowledge to be shared than as intellectual property to be protected.

At the Los Angeles conference, IBM had arranged for what today is known in the industry as a "demo" – a public demonstration – of FORTRAN. Shortly before the conference, IBM had asked a few of its customers to come up with real-world computing chores, like calculating air flows for the design of a jet wing. The problems would be coded by assembly programmers, but also written in FORTRAN. When the assembly coders were done, the two programs for each problem were put on a 704 – one hand-coded and the other FORTRAN compiled. The FORTRAN programs ran nearly as efficiently, sometimes as efficiently, as the code of the assembly programmers – that is, the FORTRAN program consumed no more machine time than the hand-coded program to solve a standard problem. FORTRAN had proved its skeptics wrong. "It was a revelation to people," Ziller said. "At that point, we knew we had something special."

Daniel McCracken first encountered FORTRAN in 1958 at New York University, where he had a graduate fellowship. By then, he had been a programmer for seven years, working mostly for General Electric. He thinks, though he is not absolutely sure, that his first FORTRAN program calculated heat flows in liquids. But McCracken does recall the sense of excitement he felt. With FORTRAN, he could suddenly program in a way that closely mirrored the mathematical problem he wanted to solve, instead of having to focus so much on coding for the machine.

FORTRAN, he figured, enabled a veteran programmer like himself to program at least five times faster than if he was working in an assembly language. "Maybe it's hindsight, but I think I also realized even then that this would open up computing to a lot more people," said McCracken, who is now a professor of computer science at the City University of New York. His instinct proved accurate, and he benefited personally from the growth in the programming population. For 20 years, McCracken was the Stephen King of how-to programming books. *A Guide to FORTRAN Programming*, published in 1961, was his first big winner, selling 300,000 copies.

The achievement of FORTRAN, perhaps most of all, was that it demonstrated that higher-level languages were possible and practical. With FORTRAN, a huge barrier fell away, opening up the software field for a steady succession of innovations over the years that would make it easier for people to program computers. Most young programmers today regard FORTRAN, if at all, as a historical curiosity. But those in computing during the 1950s still appreciate the impact of FORTRAN. John McCarthy is the creator of the influential Lisp programming language, he helped start time-shared computing in the 1960s and the field of artificial intelligence. The professor emeritus at Stanford University is not impressed by the current state of computer science research, which he finds too focused on small steps and incremental improvements instead of thinking imaginatively and aiming for breakthroughs. But when asked what really impressed him in the early days, McCarthy replied without hesitation, "FORTRAN, that impressed me."

3

The Hard Lessons of the Sixties: From Exuberance to the Realities of COBOL and the IBM 360 Project

FORTRAN HAD ALSO IMPRESSED THE IBM CUSTOMERS who saw it in 1957, and that got the attention of IBM's senior executives. Suddenly they saw this programming project, long regarded almost as an indulgence, in a very different light. Whatever this FORTRAN stuff was, it seemed to work, and customers were excited by the prospect that it would help them overcome their programming headaches. This, IBM's management understood, was a competitive advantage. FORTRAN could help IBM sell more computers. Not everyone at IBM was enthusiastic, as Frances Allen, who had joined the company in July 1957 at the age of 24, would soon learn. Her first assignment was to teach FORTRAN to the company's programmers, and it was hard duty. "There was tremendous resistance within IBM," Allen recalled. "The programmers were convinced that no higher-level language could possibly do as good a job as they could in assembly."

Over the years, the most widely-used programming languages have shared two characteristics: each has addressed a specific need in the industry at a particular time, and each made an effort to be relatively easy to learn. As the number of programmers has grown, the biggest cost in time and money associated with the adoption of a new programming language is the training cost – what economists call the "switching cost" of learning something new. FORTRAN clearly solved an industry problem, and its combination of some English words and algebraic notation gave the programming language a familiar appearance. But to programmers of the day, with their hard-earned understanding of binary and assembly languages, the "switching cost" of moving to FORTRAN seemed high. They were being asked to embrace a technology that would tend to devalue their skills. To persuade its own programmers, IBM simply issued a corporate edict that by September 1957 programming on the new IBM computer, the 704, would be done in FORTRAN. "There was a real business and marketing incentive for IBM to force its programmers to use FORTRAN," Allen recalled. "How could IBM ask customers to use FORTRAN if we weren't?"

The force-feeding worked, and with FORTRAN, a barrier dropped in computing, helping to begin an extraordinarily rapid run of experimentation and exuberance that lasted nearly a decade. There was a sense of limitless horizons. New programming languages emerged, including Lisp, Algol, COBOL, Snobol, and PL/1, among others. A new branch of research called "artificial intelligence" – using computers to try to match human intelligence – took root and gained credibility.

Software was just starting to be recognized as something distinct from hardware, and programming as a profession onto itself. But it was an embryonic profession, without real standards, qualifications, or schools that taught a body of received knowledge. This early period of unbounded optimism – perhaps even innocence – would end by the late 1960s, when it became painfully apparent that building big software systems was more difficult and costly than anticipated. The programming profession would mature somewhat, and software would be treated as a business. Yet the early years shaped the goals of computing to this day, tested the limits of the technology, and defined the basic problems.

Fran Allen, the former FORTRAN instructor, is a product of those early years of openness and daring. She rose quickly within the IBM ranks, an early assignment being to write software for the National Security Agency's Harvest supercomputer, an extraordinary machine used for communications surveillance and code-breaking during the Cold War. Allen would eventually become an IBM research fellow, and in 2001 she was working on software for the company's Blue Gene supercomputer, a next-generation project in computational biology exploring the mysteries of protein folding, one of the most fundamental processes of life. Her experience of computing in the 1960s, she recalled, made her "very optimistic about what one could do." Her views were shared by many at the time. There were predictions that computers would soon be able match human intelligence, if not exceed it, besting chess grand masters, who were seen as possessing the height of human intellect. Computing, it was said, would be a utility as common as the telephone or electricity, with every house having a console connected remotely to a digital "power plant," giving households access to everything from electronic libraries to automated shopping. It was one of America's periods of buoyant technological enthusiasm, which focused on all that computers – "big brains" – could do. "There was this enormous sense of optimism, a sense that computing could conquer all kinds of problems," Allen recalled. "The field was new and people had no firm sense of limits. It was part of the time."

Time-sharing and artificial intelligence were two of the grand computing pursuits of the 1960s, attracting funding and attention. In the late 1960s, centers for allowing ordinary people to use computers – "sharing" the time of a large machine – were sprouting up across the United States, and Wall Street was funneling money into companies like Tymshare and University Computing Company. The time-sharing startups were seen as the beginning of the computer utility industry. Yet, by the start of the 1970s, the vision of time-shared systems scattered across the nation was fading. Bigger systems – even ones that might service a couple hundred users – were proving devilishly difficult to get working. The bottleneck, once again, was software. The programming required to run time-shared systems with many users was complex, and cost far more to design and build than anyone anticipated. In the mid-1970s, the ideal of giving ordinary people access to computers would be revived, this time by the arrival of small, inexpensive hardware – desktop microcomputers, later known as personal computers.

During the 1960s, the goal of matching human-level intelligence using computers seemed tantalizingly within reach. There were predictions of computers that could see, speak, solve problems on their own, and learn from experience. Early progress was encouraging, and by the 1980s, in another wave of optimism, venture capitalists were pouring money into artificial-intelligence startups, sniffing The Next Big Thing in technology. But mimicking human intelligence in commercially useful ways was harder than expected, and the venture capitalists fled. Once more, software was the hurdle.

Both time–sharing and artificial intelligence, however, have influenced how people use computers today, and how programmers build software. Time–sharing had an impact on everything from word processors to man–machine research, which would one day lead to the point–and–click graphic icons found on personal computers today. Artificial intelligence brought new programming tools and new vistas for computing, from speech recognition to robotics. Research in the field continues to deliver advances – Deep Blue beating world chess champion Gary Kasparov in 1997, for example – even if they have come more slowly than once hoped.

During the 1960s, time–sharing and artificial intelligence research were in ascent, and one person played a central role in starting both: John McCarthy. The first public description of a time–sharing system appears to have been in June 1959, at a UNESCO conference in Paris, when British computer scientist Christopher Strachey presented a paper, "Time Sharing in Large, Fast Computers." The concept, though, had been percolating for a while before, and McCarthy had been discussing it with colleagues since the mid–1950s. "Frankly, I was surprised that not everyone regarded it as obvious," McCarthy recalled in his office at Stanford University, where he is an professor emeritus.

In the 1950s, computers handled work one job at a time, or in a sequentially–ordered group of jobs, on batches of punched cards, which became known as "batch processing." A person who wanted to use the computer typically had to write out his or her program on special paper lined with columns and rows. The coded instructions were then punched onto manila cards, with the holes denoting pieces of data to be placed somewhere or operations to be performed on data. The user went like a supplicant to the computer center and presented the cards. Each program received a meager ration of precious computer time. If there was a bug in the program, the unfortunate user had to beg for more scheduled time, which might only be available days later.

As computers became more widely used, frustration grew over waiting to get machine time. The problem surfaced first at a place like the Massachusetts Institute of Technology, which was in the vanguard of computing in the late 1950s, and where McCarthy was then an assistant professor. The MIT computer center was the first to have an operating time–shared system, a project led by Robert Fano and Fernando Corbato. The prototype system, called the Compatible Time-Sharing System, initially had just four terminals linked to a mainframe when it was demonstrated in November 1961. As Corbato observed later, "The first genesis of time-sharing began when John McCarthy wrote a key memo." In the memo to Philip Morse, director of the MIT Computation Center, dated January 1, 1959, McCarthy described time-sharing largely in terms of programming productivity, as a means to "substantially reduce the time required to get a problem solved on the machine." He thought the gain would be considerable, noting "a factor of five seems conservative." And McCarthy, never one for understatement, added, "I think the proposal points to the way all computers will be operated in the future, and we have a chance to pioneer a big step forward in the way computers are used."

In his memo, McCarthy listed the technology required of a time-shared system, including an "interrupt" feature and rapid memory allocation. For time-sharing to work, the computer had to juggle the demands of several simultaneous users, moving nimbly from one job to the next and being able to "interrupt" one job long enough to accept the keystrokes of other users. The time-sharing concept relied on new, faster machines allocating computing cycles among many users by exploiting the time humans take to think and to type instructions on a keyboard terminal.

The January 1959 memo reads in good part as a proposal for an interactive, debugging system. "The ability to check out a program immediately after writing it," McCarthy wrote to elaborate his point, "saves still more time by using the fresh memory of the programmer." McCarthy, however, also had a broader perspective on time-sharing as a technology for giving a user – any user – the illusion of having a computer of his or her own as if a personal computer. With generous funding from the Pentagon's Advanced Research Projects Agency – the patron for so much of computer science in the 1960s – MIT later embarked a vast time-sharing experiment called Multics. But McCarthy grew impatient with what he saw as the sluggish pace and the tendency of the MIT administration to view him more as a "provocative and interesting wild man," in Corbato's phrase, than as someone whose proposals should be regarded as action plans. So in 1962, McCarthy went to Stanford University to focus on his primary intellectual passion, artificial intelligence.

One of the most provocative things John McCarthy ever did was turn a phrase. He used it in a letter to the Rockefeller Foundation in 1955, seeking funds for a conference the following summer at Dartmouth College, where he was then an assistant professor in mathematics. "The phrase 'artificial intelligence' is one that I cooked up in writing the proposal to the Rockefeller Foundation," McCarthy explained. His main audience for the phrase, though, was the participants coming to the conference. He wanted them to aim high. The Dartmouth summer research project, the proposal stated, would "proceed on the basis of the conjecture that every aspect of learning or any other feature of intelligence can in principle be so precisely described that a machine can be made to simulate it." As McCarthy recalled, "My idea was to nail the flag to the mast, as it were."

McCarthy first became intrigued by the notion of using computers to simulate intelligence in September 1948, when he attended a symposium on "Cerebral Mechanisms in Behavior" at the California Institute of Technology, where he was a graduate student. At the symposium, he recalled, a few papers were presented that compared a computer to the human brain. "This comparison was somewhat limited by the fact that in September 1948 there were no stored-program computers yet," he observed dryly. But it got McCarthy thinking about how to pursue the theory of computer-simulated intelligence.

The theory dated back to 1937, when British mathematician Alan Turing described what he called a "universal machine" – a theoretical computer. He started by demonstrating that any problem that could be solved by an expert mathematician could also be solved by a clerk, given the proper instructions and limitless supplies of paper and time. The implication was that the work of any computing device, including the human brain, could be simulated by a universal machine, or computer. This is the kind of theorizing that many software designers find maddening. Well, yes, they argue, all kinds of things are theoretically possible but utterly impractical – such follies are a "Turing tar-pit," according to the engineering critique. Yet the logic of Turing's universal machine has inspired those chasing the horizons of computing, like artificial intelligence researchers, for decades. If it is theoretically possible, why not, they ask.

In 1950, Turing directly addressed the subject of machines and human intellect in his paper, "Computing Machinery and Intelligence," which began, "I propose to consider the question, Can machines think?" To artificial intelligence researchers, it is a line with all the resonance that Melville's "Call me

Ishmael" has for novelists. In his paper, Turing first articulated a test of whether a computer could simulate human intelligence: Ask a computer questions. If a human being cannot tell the difference between the machine's replies and those of another human being, the computer has attained human-level intelligence. This became known as the "Turing test." Turing's paper went on to describe what remains one of defining goals of artificial intelligence, a "learning machine." Such a computer that could "learn" from accumulated calculations – what humans call "experience" – if properly programmed so that "the rules of operation of the machine change." Turing concluded with a combination of visionary flourish and humility: "We may hope that machines will eventually compete with men in all purely intellectual fields. But which are the best ones to start with? . . . We can only see a short distance ahead, but we can see plenty there that needs to be done."

McCarthy does not think much of the so-called Turing test. That part of Turing's 1950 paper, he said, is mainly an argument with people who resisted the whole idea that machines might be able to simulate human intelligence. That is, if a person could not distinguish between the human and machine responses, McCarthy said, "then your objections are religious or something." It is unfortunate, McCarthy added, that others have converted Turing's argument into a hard-and-fast test of human-level intelligence, because it is not a very good test. "If you say that someone can't tell the machine response from a person," he said, "then the first question is who can't tell it from a person. There are people who are very readily fooled." Instead, McCarthy said Turing's major contribution was that he was "the first person who sort of had the right idea. . . . that the key to artificial intelligence was computer programs."

McCarthy soon found that the available programming tools were inadequate to the task. To pursue human-level intelligence with computers, he explained, one must first "represent facts about the world in logic." But computing at that time was mainly calculating numbers. Clearly, there was a wealth of pertinent facts about the world beyond numbers. In 1958, shortly after he became an assistant professor at MIT, McCarthy began designing a programming language that would be called Lisp. He had been working on the ideas before, but from 1958 to 1962 Lisp would be implemented and applied to artificial intelligence problems at MIT. At the time, FORTRAN was the dominant higher-level programming language, but it was tailored for the numeric problems of science and engineering. The human brain, though, is also adept at manipulating symbols – like the letters and words that form natural language, for example. For artificial intelligence, McCarthy needed a programming language that could accommodate symbolic problems as well as numeric problems. The language also had to be uncommonly flexible, so that the symbols could be manipulated freely to explore different inferences, assumptions, and new facts – just as a human might.

To create a flexible symbolic programming language – not a theoretical exercise, but a language that could be executed on a machine – was a big step. Lisp accomplished that with a combination of innovative tools, techniques, and ideas. McCarthy chose to structure his language as a series of lists of information that are then processed – Lisp stands for "list processing." All kinds of "information about the world" can be represented in lists, which then can be manipulated to make deductions and logical inferences. Lisp also liberated the programmer from some machine "housekeeping" tasks by automatically sweeping out and freeing up the computer's memory as a program runs, a feature known as "garbage collection."

McCarthy enabled Lisp programs to be represented as Lisp data – a real advance in flexibility that gave new power to the self-referential concept of the stored-program computer itself, allowing data and programs to intermingle without restraint. Lisp's design thus welcomed the process of "recursion" – the ability of a program or piece of code to summon itself, useful in solving problems that require repeated processing. Lisp could handle "deep nesting," or the positioning of programming loops within loops, enabling a program to skitter off and solve a related problem it encounters and then return to its original task. "Lisp had a freedom and flexibility that FORTRAN lacked, and Lisp bent back on itself in a way that FORTRAN wasn't set up to do," observed Guy Steele, a computer scientist at Sun Microsystems who is the author of books on Lisp and FORTRAN.

Lisp was not the first list-processing language. At the Dartmouth conference in 1956, Allen Newell and Herbert Simon, two pioneers in artificial intelligence who became professors at Carnegie Mellon University, described a list-processing language called IPL 2. But it was a language so wedded to the peculiarities of the Rand Corporation's Johnniac computer that IPL 2 was not destined for wide use. There would be other symbolic programming languages in the 1960s, including Snobol from Bell Laboratories, Comit from MIT and Formac from IBM. Yet Lisp led the way, and it demonstrated that manipulating symbols, not just numbers, was one of the basic capabilities of computing. Lisp would be an important influence on other language designers, including James Gosling, the creator of Java, the Internet programming language.

Lisp, however, never did become a mainstream programming language. It can be used for many different applications, but Lisp was not designed for a major commercial use as FORTRAN was for engineering and science, COBOL for general business data, or SQL for databases. Lisp's notation helped give it flexibility, but its appearance – a software syntax swimming in parentheses – was foreign and off-putting to many programmers. It chewed up computing cycles and ran slowly in the early days, a tarnished reputation that stuck to Lisp even as machines became faster and cheaper. And Lisp was further tarred by the commercial failure of artificial intelligence in the 1980s. Lisp is a niche language today, with a modest but dedicated following.

Marvin Minsky, together with John McCarthy, founded the artificial intelligence program at MIT, and Minsky, McCarthy, Newell, and Simon are regarded as the four "fathers" of the discipline. When asked to define artificial intelligence, Minsky once replied that "artificial intelligence labs are the places where young people go if they want to make machines do things that they don't do yet and that are hard to do with the currently existing methods. So in that sense, artificial intelligence is just the more forward-looking part of computer science." And as progress is made, the horizon keeps receding. One result is that artificial intelligence inevitably suffers from a certain inflation problem. Once a computer can do something, it is not considered so remarkable after all. A computer beats the world chess champion? OK, but look how primitive computer speech recognition still is.

Yet it is also true that artificial intelligence has proved to be far more difficult than imagined. In 1958, for example, Newell and Simon predicted that by 1970 computers would be capable of composing classical music, discovering important new mathematical theorems, and understanding and translating spoken language. Raj Reddy was one of John McCarthy's early students at Stanford University, joining the graduate program in 1963 and becoming one of the university's first recipients of a Ph.D. in computer science. Today Reddy, a professor at Carnegie Mellon, is a leading artificial intelligence researcher, particularly in speech recognition and robotics. In the 1960s, Reddy recalled,

"We thought everything was possible. We thought speech, robotics and all sorts of other problems would be solved in a matter of years. We had no idea how difficult achieving human-level intelligence would be. We grossly underestimated the complexity."

Nearly four decades later Reddy, an elder statesman in the field, observed that artificial intelligence is "really just at the end of the beginning." And in his seventies, John McCarthy, heavyset, bearded, and gray, is an irascible defender of the discipline he helped create. People who speak of the failure of artificial intelligence, he said, are mostly "ignorant journalists" – a phrase McCarthy apparently regards as a tautology – and vengeful venture capitalists. "A great part of the negative view of artificial intelligence comes from investors," he said. "They lost their shirts on it, but they were still wealthy enough to be influential." McCarthy was reared as a child of radical working-class parents, both Communist party activists, who passed on to their son the rationalist belief that technology was good for humanity. No field so embodies technological self-confidence as artificial intelligence, and McCarthy remains an optimist, albeit a grumpy one. "Human-level artificial intelligence is a difficult scientific problem and it isn't solved yet," he said. "Who knows how long it will take? But after all, it was a hundred years between Mendel and the genetic code being cracked. And nobody's claimed that genetics is done yet. And artificial intelligence isn't done yet either."

By the end of the 1950s, the computer was moving beyond science and engineering and into the heart of large companies and government organizations – on the way to becoming, as *Fortune* magazine phrased it in the mid-1960s, "the most vital tool of management introduced in this decade." Computers, it was recognized, could help manage the increasingly complicated accounting, payroll, logistics, purchasing, and manufacturing operations growing in size and reach during the postwar years. There was clearly a need to simplify programming for these management and business problems, just as FORTRAN had done for science and engineering problems. The result would be an unusual effort to fashion a programming language by committee, and an unusually successful one in that the language would become the entrenched standard for business programming. Initially, there was scattered resistance to the Common Business Oriented Language, or COBOL, from some computer companies, but resistance soon proved futile. From the outset,

COBOL got little respect from the academic community, but the criticism had no impact on its acceptance. Because it was mainly in English, COBOL was a higher-level programming language that appeared easy to learn. And it addressed a big problem of its day: how to handle business data in a programming language.

The success of COBOL, oddly, showed both that computing was becoming mainstream at the start of the 1960s and that it was still a small industry. The business of computing was sufficiently small and concentrated that a single customer could ensure that COBOL became a technology standard. The United States Department of Defense – by far the largest purchaser of computers – oversaw the design of COBOL and, after it was created, the government announced that it would not buy or lease computers unless they spoke COBOL. (In the 1970s, the Pentagon was less successful in mustering industry support for the Ada programming language, which the government sponsored and funded.)

The COBOL initiative did have the support of most computer manufacturers and users in government and industry. Two companies already had efforts underway to establish their business languages when the push for COBOL began in 1959: Sperry Rand's UNIVAC division with its Flow-Matic, a data processing language in use for years, and IBM, which was just beginning work on its Comtran, for Commercial Translator. Other computer makers did not want to see a rival possess a major new programming standard. For even if a language was openly distributed without charge, as IBM did with FORTRAN, a standard business-programming language would be a big marketing advantage for the company that created it. For computer users, the worry was that there might be a long period of confusion before a standard emerged from the messy process of market competition. Besides Sperry Rand and IBM, other companies had announced plans to develop their own languages for business data. The prospect of a clear, early winner in business languages was uncertain at best.

A collective approach to designing a business language was first suggested by Mary Hawes, a programming manager at Burroughs, a company not in the business-language foot race. She talked to Saul Gorn, a professor at the University of Pennsylvania, and they agreed that trying to develop a single business language would be in the interests of the young industry. In April 1959 a meeting was held at the University of Pennsylvania computer center, attended by representatives of a small group of customers and manufacturers, including Grace Hopper, who led the development of Flow-Matic at Sperry Rand. At the meeting, Hopper recommended that the Pentagon serve as a kind of orchestra leader for their collective undertaking. The government could provide the leadership and clout needed if a workable agreement was to be hammered out.

The group approached Charles Phillips, director of data systems research at the Defense Department, who later agreed to become chairman of the Committee on Data Systems Languages, known as Codasyl. In late May, a two-day meeting at the Pentagon was attended by roughly 40 representatives of computer makers and users. At the Pentagon gathering, there was agreement on some broad guidelines, notably the "maximum use of simple English," Phillips wrote in a report on the meeting. "We need to broaden the base of those who can state problems to computers." Shortly afterwards, there were smaller meetings of the Codasyl executive committee, which included executives from corporate users like U.S. Steel, Dupont, and Esso, in addition to a pair of advisers from the computer industry: Hopper of Sperry Rand and Robert Bemer of IBM. This body would oversee and monitor the group that did the real work of designing and detailing the COBOL language – the six men and three women initially on the "short-range committee" that first met in the last week of June 1959 and delivered the COBOL language by the end of the year. After the first meeting in June, others joined the short-range committee.

The short-range committee consisted of seasoned programmers like Jean Sammet, who began programming four years earlier, making her an elder stateswoman in the profession at the time. Like most early programmers, Sammet found her way into the field by happenstance. The daughter of two New York lawyers, Sammet showed an aptitude for math even before she learned the word mathematics. She majored in math at Mt. Holyoke College, and got a master's degree at the University of Illinois. While there, she was aware of the school's Illiac computer, but Sammet wanted nothing to do with it. To her, math was an elevated intellectual pursuit, while engineering was the grubby work of machinists – a fairly prevalent attitude among those studying math theory. "It's hard to describe how much contempt we had for the engineers in the computer center," Sammet recalled. "And I thought of a computer as some obscene piece of hardware that I wanted nothing to do with."

Back in New York, Sammet was given a teaching assistantship at Columbia University, where she studied for her doctorate. Before the school year began, Sammet took a job in the finance department at Metropolitan Life Insurance, doing calculations for dividends. Metropolitan Life was using big calculating machines in its business, and while there Sammet took a training course in preparing calculations on punched cards. "To my utter astonishment," she said, "I loved it." After Columbia, Sammet went to work at the Sperry Gyroscope Company in Long Island, and when the defense contractor decided to build a computer she became part of the team. In early 1955, Sammet recalled, her boss, Arthur Hauser, asked her if she wanted to be a programmer. Sammet replied, "What's a programmer?" Hauser said he did not know, but he knew the computer project would require a programmer. Was it anything like dealing with punched-card equipment, Sammet inquired. Her boss said he suspected so. "That's how I became a programmer," Sammet explained, smiling at the memory.

The early Sperry computer, called Speedac, was mainly a learning experience for the company and everyone involved, including the programmer. The company knew it had to embrace digital computers, if only because they were becoming crucial tools in the defense industry. Sperry went further in 1955, entering the computer industry itself by purchasing Remington Rand, whose UNIVAC machines had made it the early leader in commercial computing. For Sammet, the merger meant ready access to the powerful UNIVAC computers. She and a couple of colleagues would regularly take the evening train down to the UNIVAC division in Philadelphia so she could run programs at night on the new machines before they were shipped to customers. "We were the beta testers of the time, alpha testers really," Sammet observed. The practice was encouraged by the ranking UNIVAC manager, Grace Hopper.

Sammet also began to teach programming in night classes at Adelphi College. Her first year of teaching, starting in the fall of 1956, was done without a text. By her second year, Dan McCracken's first book on programming had been published, and she used it in class. In 1957, Fortran was released and she taught the new programming language, reading hurriedly in the IBM manual to stay a step or two ahead of her students. "I knew as much about FORTRAN as that potted plant," said Sammet, gesturing toward the ornamental foliage in the lobby of her apartment building in Bethesda, Maryland. Such blunt-spoken candor would be a Sammet trademark throughout her career, and she would also continue to teach part-time, including a lecture in Boston where one of her pupils was Dennis Ritchie, who later created the C programming language.

Over the years, Sammet would teach many programmers in the classroom and on the job, and she found that predicting who would make a good programmer was not easy. With her fondness for precision, logic, and math, Sammet had the knack. An ability in math, she noted, tended to be a good indicator, but by no means foolproof. She recalled training two programmers side by side, one a young woman with a home economics degree and the other a man with a graduate degree in applied math. The home economics major was a "crackerjack programmer," she recalled, while the mathematician was a dud. "A person either has or has not the mental characteristics that makes someone a good programmer, and it has little to do with educational background," Sammet observed. In recruiting, she looked for a certain intellectual passion, a willingness to get lost inside the code and the machine for sheer fun of it. At Sylvania Electric Products, which she joined in 1958, Sammet had a habit of asking job candidates whether they were more interested in doing the "useful" programming of making software applications that perform business and scientific tasks or the "useless" programming of developing systems software – compilers, device controllers and other programs that control the working innards of the computer. "If the answer was useful, I told them, 'I don't want you,' " she said, recalling her half-joking reply.

Sammet was assigned to the near-term COBOL committee, whose mandate was to "recommend a short-range composite approach (good for at least the next year or two) to a common business language for programming digital computers." The "composite approach" referred to essentially blending the work of three existing business-language projects: Sperry Rand's Flow-Matic, IBM's Commercial Translator, and an Air Force-led version called Aimaco, which borrowed heavily from Flow-Matic. The common business language of the short-term committee was to be "machine-independent," meaning it would be a higher-level programming language under which each computer maker would then write its own machine-specific compiler. In addition, there was an intermediate-range committee and plans for a long-range committee. In fact, the intermediate-range committee never did produce a language, and the long-range committee was never formed. The short-range committee called itself the PDQ ("pretty damn quick") committee, even in its written reports. "Most of us viewed our work," Sammet wrote later, "as a stop-gap measure – a very important stopgap indeed, but not something intended for longevity."

An initial deadline of three months was set for the work of the committee, but that was soon recognized as unrealistic. They would be given until December 1959, nearly six months, to finish. Besides the technical specifications, there was the matter of naming the new language. At a meeting on September 17, names were discussed. Six were suggested including Busy (Business System), Infosyl (Information System Language), Datasyl (Data System Language) and Cocosyl (Common Computer Systems Language). None seemed quite right, and no agreement was reached. Though not members of the short-range team, Grace Hopper and Robert Bemer, the advisers to the supervisory Codasyl executive committee, kept a watchful eye on their work. Hopper and Bemer discussed the name issue that evening. "Nobody could come up with anything worth a damn," Bemer recalled. So Bemer, who is fond of acronyms and thought up Codasyl, says he suggested COBOL. "Grace said, 'OK, sounds good to me,' " Bemer recalled. "There was no real debate about it." The next day, COBOL was agreed to as a catchy contraction of Common Business Oriented Language, and apparently not a lot more thought was given to a name that would become one of the indelible terms of computing. (In fact, in a published reflection on the programming language in 1971, Bemer wrote, "We can't find a single individual who admits coining the acronym 'COBOL.'" Later Hopper, who died in 1992, credited Bemer with thinking up the name. Hearing Hopper tell the anecdote, Bemer says, jogged his memory. Told of Bemer's account, Sammet says she has her doubts.)

By October 1959, the short-range team had made some progress, but they were getting bogged down in lengthy debates. Besides the original nine members of the committee, other industry representatives had been encouraged to join. The meetings were typically attended by 12 to 20 people. These were unwieldy forums in which to try to hammer out the technical specifications for a programming language. "There is a real limit to what can be done by committee," Sammet recalled. "People just sitting around and talking wasn't going to get it done." It was decided that a six-person group would go off on its own and try to come up with detailed specifications for the language. This smaller group, which included Sammet, really created COBOL. Much of their work was done in New York City, where the six were holed up in the Sherry Netherland Hotel for two straight weeks, toiling round-the-clock toward the end. That small team produced their draft report in early November. Then, for five days, their draft was reviewed and debated in front of the short-range committee. Some changes were suggested, but mostly their work remained intact.

As a programming language, COBOL is an odd-looking beast. If using the fairly large subset of English words included in COBOL, a person can just write out what appear to be natural-language statements:

MULTIPLY HOURLY RATE BY HOURS WORKED GIVING GROSS-PAY.

Programming purists often regard COBOL as a blight on the landscape of computing. But there were no programming language theorists involved in the design of COBOL. Instead, COBOL's creators were a small group of commercial programmers, working under extraordinary time pressure. Their marching orders were mainly to cobble together ingredients from a couple of early experiments in data-systems languages and to use as much English as possible. The idea was to make programming accessible to nonprofessionals, and to make business computing comprehensible – and thus less intimidating – to the management class. Even within the group that created COBOL, however, there were real differences over how simple, or simple-minded, a programming language for business should be.

The issue came to a head in the debate over "arithmetic verbs." One group believed that business users should not be forced to grapple with math formulas, so the language should include the arithmetic verbs ADD, SUBTRACT, MULTIPLY, and DIVIDE. Others believed that this was a silly, dumbing-down of the language, and that anyone working with computers would be comfortable with simple arithmetic formulas. The dispute was settled by compromise. The arithmetic verbs went in, but formulas were also permitted in COBOL. A formula statement, however, had to be preceded by the verb COMPUTE. "The biggest argument was over arithmetic verbs," Sammet said, "because that issue brought the religious divisions to the surface."

COBOL did add an important improvement to the field of programming languages, in the realm of how data is described and represented. And that, in turn, contributed to the development of database technology. COBOL, for example, included a feature called "picture clause," which enabled data to be described in hierarchical tables. This facility was included to help organize basic business or customer data in a program: names, addresses, Social Security numbers, phone numbers, and the like. To create a "picture" of the data, the programmer uses a set of defined numbers or characters in a visual pattern. For example, the digit "9" is used to represent a corresponding number in the same position in the "picture." So the raw data of a phone number, stored in the computer, might be: 2125561234. It could be filtered through a picture clause – (999) 999–9999 – and then appear on a user's screen as (212) 556-1234. "Cobol did hierarchical data layouts very well, and that wasn't even a gleam in the eye of FORTRAN or Algol," said Brian Kernighan, a professor at Princeton University.

COBOL fell short in its implied promise of "programming in English." The notion had undeniable appeal and its greatest champion and promoter was Grace Hopper. She would become software's leading stateswoman and a rear admiral in the Navy. A tiny, tireless, and plain-spoken woman, Hopper was a technology evangelist whose vision of making software more accessible prodded the industry in that direction. She spoke of this baffling technology in stories and anecdotes that were understandable to people outside of computing. And she was no stranger to the simplifying exaggeration.

Her story-telling flair, for example, would ensure that the term computer "bug" would be permanently linked to her. Repeatedly, she told an anecdote about using a tweezers to pull a dead moth out of the Harvard Mark II computer on Sept. 9, 1947. "From then on," Hopper said in a 1981 speech, "when anything went wrong with a computer, we said it had bugs in it." Yet the term "bug" referring to glitches in machinery goes back at least to the late nineteenth century, including Thomas Edison referring to finding a "bug" in his phonograph in 1889. And Hopper's papers in the Smithsonian Institution's National Museum of American History show that the term bug had been used by Hopper and others to describe computer problems years before 1947. But the moth story was the one that stuck.

In the case of COBOL, Hopper was a prime mover behind the project. At

Sperry Rand, Hopper led the development of Flow-Matic, which contributed much to COBOL. She was an advisor to the Codasyl executive committee, which oversaw the creation of COBOL. And Hopper traveled around the country, stumping for COBOL to persuade companies to adopt it. She would become widely known as the creator, inventor, or "mother" of COBOL, but in fact Hopper was not directly engaged in the design and technical specification of COBOL. "The main incorrect statement about COBOL is that Grace Hopper invented it," said Sammet. "She never walked into any committee meeting room that I was in . . . I yield to no one in my admiration for Grace. But she was not the mother, creator or developer of COBOL."

When asked, Hopper was straightforward, even precise, about her role in the COBOL effort. In 1976, computer-book author Dan McCracken told Hopper he was going to dedicate his book, *A Simplified Guide to Structured COBOL Programming,* to her and call her "the mother of COBOL." No, Hopper replied, a better description would be to say she was the "grandmother of COBOL," which seems accurate, given the Flow-Matic heritage. "Grace was modest in her way," McCracken observed.

She was a consummate saleswoman, though. McCracken recalled attending a lecture on COBOL that Hopper delivered in Philadelphia. She emphasized the English language programming, and it worked splendidly as long as one kept to COBOL's stinted English vocabulary and to a select group of problems. It was misleading, McCracken said, because presenting COBOL as allowing a person to program in English gave the mistaken impression that here was a programming language with the flexibility of a natural language, able to accommodate human ambiguity and imprecision. But a period out of place, or some other errant keystroke, could crash a program in COBOL as surely as in assembly language. "COBOL was just as fussy and unforgiving as any other programming language," McCracken observed.

Business executives never really had any inclination to read the COBOL code written by their technical staff. Professional programmers soon adopted mnemonic abbreviations for their code, shunning the verbosity of writing out the English words as if sentences. "Grace was misguided about how the English would be used in programming," McCracken said. "It just didn't work out the way she had expected." But she was right, he added, about the big ideas of the necessity of simplifying programming and the need for a language for handling business data. McCracken compared Hopper in very broad terms to Columbus, who sailed West and did not find the Spice Islands, but did find a New World – extremely significant, if not the original goal. "COBOL was the business FORTRAN," he said. "It was a huge accomplishment, and Grace Hopper led the way."

The rise of "software engineering" was driven by the same force that led to COBOL – the recognition that computing was moving into the mainstream of business, commerce, and government, and that software was crucial to that happening, but also a growing problem. IBM seemed to be the company best-placed to address the software engineering problem, and it was T.Vincent Learson, the second-in-command to Thomas Watson Jr., who pressed the issue. Learson, six foot six, angular and imposing, was Watson's field general and a gruff man of action. IBM's research scientists and technical staff regarded Learson as a highly intelligent salesman and executive – a different animal, but a good one. In 1960, Learson appointed David Sayre, a veteran of the original FORTRAN team, to a new corporate staff position, director of programming. Sayre organized a conference of IBM programmers to suggest ways to improve the company's software development.

More than 100 IBM programmers – the "cream of the existing crop," Sayre noted – gathered for a week-long conference at the Bald Peak Colony Club in New Hampshire in June 1961. But the Bald Peak gathering was to be more than a discussion forum for programmers. Watson, Learson, and the presidents of the sales and product divisions also attended. What emerged was an "Action Program" with 35 steps to be taken. The steps all flowed, Sayre observed, from the "recognition that the product IBM put on the market was half software and half hardware," and the recommendations included items ranging from salaries to software testing. Software, in theory, was to be given equal status with hardware. That, it seemed, would mean treating hardware and software the same, even using similar terminology. IBM, for example, had an established engineering process for the design and production of hardware. It was divided into three clearly delineated steps, A, B and C, from conceptual design to a finished machine, fully tested, ready to plug in and ship to a customer. The IBM programmers adopted the parallel terms "alpha," "beta" and "gamma" for software. Indeed, alpha and beta survive to this day as terms widely used in the industry, but mainly to describe unfinished programs thrust out on users to help with the endless chore of debugging.

The desire to treat software more like its hardware sibling began in earnest in the 1960s. Equal treatment was one motivation, at least at the start, but it was also an effort to make programming a more structured discipline like hardware engineering. Looking back, Sayre observed that the Bald Peak conference had the regrettable effect of beginning a management mentality that served to "rigidify" programming at IBM. In the early days, he recalled that programming at IBM had an informal Silicon Valley flavor, which tended to encourage innovation. By the late 1960s, the working environment had become much more regimented. The disciplines of hardware engineering fit uneasily in the more ethereal realm of software. "Software is a much more plastic object than hardware," Sayre said. "You whip it up, squeeze it, and you can dream."

Yet the impulse to try to make software development a more disciplined – a more engineered – process was logical and inevitable, given the immense problems of building large software programs that surfaced during the 1960s. Advances in storage and circuitry had raced ahead, making computers faster and more powerful. And a 10-fold increase each in memory and processing speed combined to deliver a 100-fold improvement in system performance, which was achieved in the first half of the 1960s. Yet if the machines were suddenly capable of scaling new heights, the gains in software came grudgingly. Trying to move from a program with 10,000 lines of code to a program with 1 million lines opened a Pandora's Box of software woes. The increase in complexity was more biological than mechanical – all that code that somehow had to fit together and get along. And adding more people to big jobs often merely added to the confusion, instead of lightening the labor for all. Building software was not like building a bridge or a road. Programs did not *scale up* according to the orderly rules of physical things. Big software projects were invariably delivered late and over budget, and the programs were often unreliable.

The signature software project of the 1960s was the operating system for IBM's 360 line of mainframe computers. The OS/360 was a bold, ambitious design with bright ideas and bad ones, a sprawling smorgasbord of software. An operating system is often described as the command and control system of a computer. The best-known operating systems today are Microsoft's Windows, Apple's Macintosh and the Unix operating system, whose versions range from Sun Microsystem's Solaris to the freely distributed Linux. By the early 1960s, most computer manufacturers were providing basic operating systems in response to demand from customers, who were using computers for an ever-wider range of tasks. They wanted their programmers to focus more on their scientific and business problems, and less on the inner workings of the computers themselves. So an operating system, supplied by the manufacturer, would orchestrate the flow of work through the machine.

The OS/360 was a daring expansion of the concept of an operating system of the early 1960s. It was very nearly an unqualified disaster for IBM, but the OS/360 was eventually redeemed by its good ideas, IBM's deep pockets, and the Herculean labor of an army of embattled programmers. The 360 mainframe line proved to be an enormous success for IBM, ensuring the company's dominance of the industry into the 1980s. And the OS/360 became an industry-standard technology, the Windows of its day. And even today, much of the industrial-strength corporate and government computing of the world is done on mainframe computers running software that has been updated time and again, but is a direct descendant of the OS/360.

The 360 line, announced in April 1964, was a family of six machines – its name referring to the 360-degree, full circle of computing. The product line was conceived with marketing and the customer's software in mind. The different-sized 360 computers would be "compatible," meaning that the programs written for one machine would run on others in the line. So, as a company grew, its investment in scientific, accounting, or manufacturing software applications would be protected as the customer moved to larger machines in the 360 family. The 360 included "emulation" software so that programs written for previous IBM machines could be switched over the 360. The company also promised that the OS/360 would be capable of "multiprogramming," enabling the computer to run several programs at once. It was a great marketing pitch, but delivering on the grand promises would be excruciating and expensive.

In 1966, speaking to a group of customers, Watson acknowledged the software travails of the 360 with gallows humor, noting that the annual programming budget appeared to be escalating rapidly. "A few months ago the bill for 1966 was going to be $40 million," Watson told them. "I asked Vin Learson last night what he thought it would be for 1966 and he said $50 million. Twenty-four hours later I met Watts Humphrey, who is in charge of programming production, in the hall here and said, 'Is this figure about right? Can

I use it?' He said it's going to be $60 million. You can see that if I keep asking questions we won't pay a dividend this year."

Watts Humphrey took over as IBM's director of programming in January 1966. In late 1965 a few people, including Humphrey, had been asked to submit brief papers on how IBM might better manage its programming corps. His paper, Humphrey recalled, was a "tirade on the need for disciplined planning, scheduling and tracking." With the effort to produce the OS/360 delayed and in disarray, such a tirade was precisely what IBM's top management wanted to hear. Within days, Learson told Humphrey, who was 39, that he was in charge of the company's programming.

As a child, Humphrey was familiar with both privilege and the discipline needed to overcome a learning handicap. His grandfather had been the first president of the New York Federal Reserve Bank and his father, Watts Humphrey II, was the treasurer a large insurance company in New York. But young Watts flunked out of first grade. It was a time, in the early 1930s, when many children with dyslexia were just considered to be "slow." Humphrey's parents, however, had their son examined by experts, who made the diagnosis. To deal with his problem, the family moved to Connecticut so Humphrey could attend a new private school for children with special learning difficulties, The Forman School, where he was individually tutored until he was nine years old.

Once he had learned to cope with his dyslexia, Humphrey caught up quickly, graduating from high school at 16. He excelled in math and science, and to this day is grateful to the teacher, John Yarnelle, who made those subjects come alive for him ("he changed my life"). Humphrey got accepted as a scholarship student at the California Institute of Technology, but he chose not to go, enlisting in the Navy in 1944 instead. After the war, Humphrey went to the University of Chicago, where he majored in physics. He went on to get an MBA at Chicago, supplementing his business studies by taking electrical engineering courses at the Illinois Institute of Technology.

In 1953, Humphrey joined Sylvania Electric Products and was put in charge of a small engineering group in Boston, which was making a special-purpose cryptographic computer for the government. To burnish his skills, Humphrey took a summer course in computer programming at MIT. It was taught by Maurice Wilkes and a few Cambridge University colleagues who had built the first stored-program computer. Humphrey's introduction to programming was writing a simple demonstration program to produce a bouncing digital "ball" on a cathode ray screen. It was crude programming, he recalled, but being able to stir the machine to perform even an elementary task was a thrill. The machine, the Pentagon-funded Whirlwind, was very fast at the time, with a processing speed of a megahertz (inexpensive personal computers in the spring of 2001 were 800 times faster).

In 1959, Humphrey joined IBM after a friend of his father's arranged an interview with Thomas Watson Jr. himself. Humphrey proceeded to manage a variety of engineering projects of increasing size and importance to the company, until the OS/360 mess was handed to him in 1966. His first week on the job proved sobering. The marketing division had just put out a "Blue Letter" on the timing, pricing, and programming support for the IBM 360 model 91, the largest machine in the line. The so-called Blue Letters, which were sent to the IBM sales force, were effectively public announcements, because the marketing group passed the information along to customers and the industry. Humphrey went to the Poughkeepsie lab, where much of the programming for the OS/360 was being done, and asked how the model 91 work was coming along. To his dismay, Humphrey discovered there was "not only no plan, the project was not even staffed or funded." He immediately called Jack Rodgers, vice president of marketing, and recalled starting the conversation facetiously by saying, "Jack, I didn't know you were a programmer." He demanded that a corrective announcement be put out. The Model 91 would have to wait, because no one had written its software.

Next, Humphrey toured the company's software development labs, only to find they were "operating in a state of near panic." They were slipping so far behind schedule that, Humphrey recalled, "everything but the immediate crisis got deferred," an approach that served to make matters worse in the long run. There were no plans. By then, the software shipment deadlines had been pushed back three times. The work on OS/360 was roughly a year behind schedule, and things appeared on the verge of getting out of control altogether. Drastic action was required, and Humphrey took it. He went to see Frank Cary, the senior vice president for development and manufacturing, and told him that "since all the delivery dates weren't worth anything anyway, I intended to cancel them." He then gave all the software managers 60 days to come up with detailed development plans for their projects. With the plans in hand, Humphrey proceeded to add 90 days to each one.

How had things gotten so out of hand at a company held up as a paragon of management efficiency? The challenge of the 360 was going to be daunting, if only because project was such a technological stretch. As one IBM executive told a writer for *Fortune*: "We are trying to schedule inventions, which is a dangerous thing to do in a committed project." In retrospect, the design could have been simplified, eliminating features and functions that were redundant or overlapping. Yet the project was so vast that "recognizing the overlap in your mind in the heat of battle was very difficult," observed Frederick P. Brooks Jr., who led the project until 1965, when he left IBM for the University of North Carolina, where he had previously accepted the post of chairman of its new computer science department.

The difficulties were compounded by the special challenges of software design and development. The hard lessons of the OS/360 saga were presented with uncommon clarity and wit in *The Mythical Man-Month* by Brooks himself. In his classic work on building software, first published in 1975, Brooks warned of the "second-system effect" – the understandable but perilous tendency of engineers to abandon self-discipline the second time around. "An architect's first work is apt to be spare and clean," he wrote. "He knows he doesn't know what he's doing, so he does it carefully and with great restraint." After the restrained success, the second system tends be a far grander conception. "The second is the most dangerous system a man ever designs," wrote Brooks, noting that the OS/360 was the second system for most of its designers.

Yet it is also true that the second-system syndrome could apply to building just about anything – a bridge, a house, a piece of furniture. Where Brooks made his real contribution was in distilling certain principles that, if not unique to software, applied mainly to the design and construction of complex, dynamic systems like large computer programs. The best-known of these principles, Brooks's Law, states: "Adding manpower to a late software project makes it later." Software, he was saying, was not really like building some vast physical edifice like a baseball stadium or some ancient monument – just add more laborers and the job will go faster. No, software was different.

To explain, Brooks relied on his own hard experience with the OS/360 and the work of others in the then-growing field of software engineering. The problem, he said, began with the fact that, as studies showed, some programmers were 10 times as productive as their peers. (Some computer scientists have argued that the difference between run-of-the-mill and great programmers is more like a 100-fold.) Regardless of the multiplier, the import of the observation is the same: in programming, a profession with elements of artistry and creativity, workers are not interchangeable units of labor.

So the question becomes how to organize a big software project most efficiently. Never easy, Brooks advised, but he decided that the best answer revolved around "surgical teams." The concept, he noted, was borrowed from Harlan Mills, a former IBM manager and computer scientist who wrote extensively on "software productivity." Software projects, Brooks explained, were best organized as a collection of small, focused groups. Each group, he added, should be "organized like a surgical team rather than a hog-butchering team. That is, instead of each member cutting away on the problem, one does the cutting and the others give him every support that will enhance his effectiveness and productivity." This results in a hierarchical team structure with a "chief programmer" supported by various classes of assistants from an "alter ego" down to testers, editors, and secretaries. The intent is to maximize the effectiveness of the most effective person – the chief programmer. The surgical team structure, Brooks said, would best allow the chief programmer to focus on big, overarching issues of design while handing off matters of implementation and administration to others. On a big project, the surgical team organization would also greatly simplify communication – a significant "cost" in terms of time, energy, and mistakes on a big project – because the key communications would occur among the chiefs, instead of everyone else.

By erasing all the delivery dates in 1966 and ordering the programming groups on the OS/360 to draw up credible plans, Humphrey was putting IBM's software managers on the spot, but he was also finally giving them a voice. The machine's design had come first in the project, IBM being first and foremost a hardware company. Software was referred to as "programming support" for the machines. The hardware myopia was hardly surprising. IBM did not sell software separately from its computers until 1968, when government pressure and market forces prompted the company to "unbundle" software and hardware. IBM's top management did not really understand programming, and too often set the programming schedules according to the marketing plans. "A lot of the problems came from management," noted John H.

Palmer, a former IBM programming manager and coauthor of an official history of the 360. "Programmers should have had more freedom to set the schedules."

With the plans in hand, Humphrey began announcing a series of releases for the OS/360 spread out over two years. The company met those deadlines one by one. IBM eventually delivered on its promise, though a year later and at four times the initially budgeted cost – roughly $500 million instead of $125 million. Reliability and performance problems, however, would persist, so that the programming corps remained under intense pressure to do emergency debugging and maintenance work through 1968. Measuring the impact of an executive is always uncertain, though the fact remains that the OS/360 ship steadied under Humphrey's watch.

But Humphrey had the time necessary to work through IBM's software troubles because of a shrewd decision by the 360's designers. They included "emulator" technology – special programming and circuitry – that enabled customers to run software they had developed for IBM's previous 7000 and 1400 series machines on the 360 machines. As long as they could run their existing applications, customers purchased 360 computers and were willing to wait for the OS/360, instead of switching to buy computers from other manufacturers. The emulator technology bought IBM the time it needed to finally get the OS/360 in shape. "That's what saved our bacon," Humphrey said.

The IBM 360 became one of the great business success stories of the postwar era, a "bet the company" gamble that paid off handsomely. The 360 line, along with its successor the 370, defined the mainframe era of computing. Yet the years working on the OS/360, from 1963 to 1968, were exhausting and often demoralizing for IBM's programmers. "The hours were long, and they were under constant pressure from management," observed Palmer, the IBM historian. "Many programmers felt that they gave their best years to the 360 with little to show for it. People came out of it discouraged and dispirited, and many left."

The 360 software morass was a significant contribution to the mounting evidence during the 1960s that big software programs were extremely difficult to build effectively and reliably. Unforeseen problems were continually surfacing in large programs, sometimes with dire consequences. In 1962, for example, the United States Mariner I spacecraft, on a mission to Venus, had to be destroyed shortly after leaving the launch pad, after it swerved out of control. Investigation found that the problem was caused by a single errant character in a FORTRAN program that was part of the rocket's guidance system.

In 1968, NATO sponsored a conference prompted by concerns that the existing "software crisis" posed a threat to the economic health and military readiness of the West. Papers were presented and speeches given, but mostly, it seemed, everyone agreed that a long slog lay ahead. No quick answers or "breakthroughs" would solve the software engineering problem. "There was a 'crisis' in the software field," Bernard A. Galler, a computer scientist at the University of Michigan, recalled later. "Everyone was aware that hardware was moving ahead rapidly, and we could see the potential use of computers, but we couldn't see how to exploit the new technology adequately with the 'craft mentality' that we all had in generating software. . . . How should one manage a team of hundreds of people, if each is crafting a beautiful piece of code which would have to interact with many other such individual creations?"

Watts Humphrey emerged from his IBM experience in the 1960s and afterwards with a deep belief that software can be made better and more reliable by the intelligent application of engineering disciplines. In 1986, he joined the Software Engineering Institute at Carnegie Mellon University, and he has been crusading for improved software quality ever since. In 2000, India – which is trying to gain a competitive advantage in software with superior quality, just as Japan did in automobiles in the 1980s – named a research institution in his honor, the Watts Humphrey Software Quality Institute. To Humphrey, the trouble with software starts with the job itself. The hapless programmer, he said, is required to take a fuzzy human problem and "reduce it to absolutely precise description so it can be executed on a machine. That is not a human activity."

So, inevitably, building quality software – meaning virtually bug-free – is a constant struggle. Lousy software, Humphrey explained, may have 10 defects per 1,000 lines of code. It probably works, but the user will likely encounter trouble. Most commercial software today is somewhat better, perhaps 5 errors per 1,000 lines, which still is not very good, according to Humphrey. Make the reasonable assumption, he continued, that 1,000 lines of code have an average of 17 characters a line, or a total of 17,000 characters. So the "lousy" code has 10 mistakes in 17,000 keystrokes. "That's lousy for software," Humphrey said. "But it is an extraordinary performance in most any other field of human endeavor. Programming is very, very disciplined work." The software field, Humphrey says, can be divided into two sides: the big innovative bursts of artistry, which he called "monument creation," and hard steady work of engineering, which is refining, maintaining, and testing software. "For the most part, 90 to 95 percent, we're in the engineering business, not the monument creation business," he said.

The precise percentages can be debated, but there is undoubtedly a lot to Humphrey's point. Still, it is not the whole point. It may be only 5 or 10 percent, but path-breaking creativity over the years has defined the field of software and pushed it forward. Without the 5 or 10 percent of genuine innovation to build upon, why bother with the engineering processes? And the best tools used by everyone in software – programming languages and others – are the work of the monument builders. In 1993, Fred Brooks, the software engineering guru, spoke at a conference on the history of programming languages sponsored by the Association for Computing Machinery. His subject was software design, and at one point he noted that some programming languages and operating systems have "fan clubs" – which he defined as "fanatics" – while others do not. "What are successful, widely used, effective, valuable contributing products," he asked, "that I never saw any fan club for?" He listed a handful including COBOL, OS/360, and Microsoft's Disk Operating System. On the other side, his list of software with ardent fans included FORTRAN, Pascal, C, Unix, and the Macintosh operating system.

The difference, Brooks said, was that the languages and operating systems with fanatical fan clubs were "originally designed to satisfy a designer or a very small group of designers." Whereas the successful products unable to inspire fan clubs, he noted, were "designed to satisfy a large set of requirements" – they were "done inside of product processes." So, Brooks asked, "What does that tell us about product processes?" His answer: "They produce serviceable things, but not great things."

4

Breaking Big Iron's Grip: Unix and C

KEN THOMPSON COULD NOT WAIT TO GET HIS HANDS on the programmers' manual for the IBM 360 mainframe. Since the company had announced the new line the first week of April 1964, Thompson had been on the phone daily, asking the IBM office in San Jose for a copy. The day a programmers' guide was available, he hopped in his dark blue 1959 Volkswagen – a classic Beetle – and sped down the Nimitz Freeway from Berkeley to San Jose.

Thompson, a student at the University of California at Berkeley, was not rooting for IBM on his drive to San Jose, which is scarcely surprising given the time and place. A few months later, the Free Speech Movement took to the streets in Berkeley, protesting a university administration they found "repressive." A year after the assassination of John F. Kennedy, college students across the nation were becoming more aware of America's problems of race and violence, and Berkeley was leading the way. In the post-Sputnik space race, university curriculums were being steered by government and corporations more toward math and science – becoming more disciplined and more regimented, it seemed, to the student leaders. In the language of dissent, "the system" and "the machine" were terms of unqualified evil. A typical leaflet announcing a demonstration declared, "Except to threaten and harm us, the machine of the administration ignores us. We will stop the machine." (Though all was not grim earnestness. The leaflet ended: "Come to the noon rally. Joan Baez will be there.")

By 1964, IBM was seen as a corporate symbol of the authoritarian tech-nocracy that many students so despised. Thompson, however, had a more par-ticular gripe with the company. The engineering student's rebellion manifested itself not in the streets, but in the university's computer center – his natural habitat, a programmer's playground for someone with his skills. Thompson's disdain for IBM was in good part a matter of technology culture, or what he viewed as the company's lack of it. The company marketed its "beautiful blue boxes," he said with sarcasm years later, to business executives who were wined and dined by the buttoned-down, wing-tipped warriors of the IBM sales force. Thompson found the IBM marketing "vulgar" and the company's computer designs too often "completely ignored the people who had to deal with their machines" – the engineers and programmers.

By then, Thompson had experience not only with IBM mainframes, but also with the emerging alternative in computing represented by the minicomput-ers of Digital Equipment. The company had not yet hit its stride, but even the early models of Digital's PDP series opened a window on a different style of computing. These machines were lower in cost, smaller in size, and spoke of a far different culture than the mainframe, with its glass-enclosed air-conditioned rooms patrolled by specially trained "operators." The PDP minicomputer was small, open, and inviting in comparison. Access was not restricted by corporate vetting or rank. Minicomputers were adopted first in research, engineering, and academic settings, and their effect was to lower the cost and barriers to exper-imenting with computers for the curious. Young researchers and students could get their hands on the machine themselves. To them, the minicomputer was in tune both with the times and their hackers' approach to computing. It seemed to embody the more informal, tinkering culture of the lab work bench, while IBM represented the regimented culture of the accountant's ledger.

After picking up a copy of the 360 manual, Thompson placed it beside him and read as he drove from San Jose through Fremont, Hayward, Oakland, and back to Berkeley. Not a recommended driving practice, but he could not help himself. As he read, Thompson felt sure that here was proof that IBM's hubris and disregard for the programmers and engineers – the people who mattered most – was carrying the company toward its comeuppance. The scheme for handling data addresses in memory was nightmarishly complex, and the 360 operating system was "too big and slow, too complicated and too baroque," he recalled. It was like a ship overburdened with cargo and encrusted with barnacles – a bankrupt design driven solely by the dreams of IBM's sales force, straining to satisfy all manner of customers to embrace the largest possible market.

"I almost got into an accident, I was so excited," Thompson recalled in the fall of 2000. "'That's it,' I thought. This is the machine that is going to take IBM down the toilet. I was sure it was going to be the Edsel of the computer business."

Thompson chuckled at the memory, sipping coffee, seated on a worn couch in a makeshift meeting room off the computer science laboratory at Bell Labs in suburban New Jersey. Tall and husky, Thompson was dressed in jeans, running shoes, and a work shirt. His hair fell to his shoulders from a receded hairline, and with his long salt-and-pepper beard and penetrating gaze, there was some truth in a colleague's comment that Thompson resembled a "well-fed Rasputin."

In one sense, Thompson's youthful assessment of the IBM 360 was right on target. The burden the 360 placed on programmers was daunting, and its operating system was a monster to produce and maintain. But Thompson was reading the programmers' manual with the elitist purity of a software artist and builder. The IBM design, with all its technical tradeoffs, was deemed necessary to attain the company's goal of developing a line of machines intended to cover the 360-degree "full circle of computing," enabling customers to move up the line to a larger machine in the 360 family without having to rewrite the software for its accounting, personnel, and manufacturing.

Thompson was mistaken about how the signature machine of the mainframe era would fare in the marketplace, yet he was right about the shortcomings of mainframe computing. In 1969 Thompson, working closely with his long-time Bell Labs collaborator Dennis Ritchie, created the Unix operating system. Lean, simple, and elegant, Unix was the antithesis of the IBM 360 operating system. Unix began simply as a tool for the computer scientists at Bell Labs – software built by programmers, for programmers. It had no business plan or marketing strategy, but gradually Unix moved out of Bell Labs into the university and engineering community, where it flourished. Engineers, researchers, and computer scientists loved Unix because it was an ideal medium for experimentation – an operating system composed of software tools that could easily be linked together, as if screwing together lengths of "garden hose," in the memorable phrase of one Bell Labs scientist.

Unix was an operating system made to lead the revolution against the mainframe – decentralized software designed as mix-and-match tools for the decentralization of computing. Time-sharing, to be sure, was a step toward a more personalized style of computing – an illusion of individual control, as software divvied up the resources of a central machine. Beginning in the mid-1970s, and then becoming a major force in the 1980s, the personal computer would go the furthest in putting computing power in people's hands – one machine for one person. Yet the bridge between the mainframe and the personal computer was the minicomputer, which in the early 1970s became inexpensive enough so that they spread across academia and business. Suddenly, purchasing a computer was no longer a multimillion-dollar decision that required a nod from the president's office. And for its first decade, Unix rode the minicomputer.

Unix was a technology that engendered a philosophy. Initially designed for a time-sharing computer, Unix was intended not merely as a programming environment but, as Dennis Ritchie described it, "a system around which a fellowship could form." This notion of communal computing was a direct forerunner of what in the 1990s would become the "open source movement," in which programmers around the world share code and collaborate on software projects. It is no accident that the poster child for open source is a Unix clone, Linux, nor that most of the server computers that act as the data-serving hubs on the Internet run some variant of Unix.

It seems no small irony that Unix – a project born of elitism – software by programmers for programmers, with no thought to users other than colleagues at Bell Labs, would become the touchstone of communal computing. The same culture, and people, also developed one of the most popular computer languages: C, which Ritchie created with help from Thompson specifically to implement Unix. And Ken Thompson was the creative fulcrum – the source of most of the big ideas, a programmer without equal. His colleague and former boss M. Douglas McIlroy observed that during "those years when everything turned to gold" Thompson was "the chief alchemist."

Ken Thompson was born in New Orleans in 1943, but he didn't stay long. His father, Lewis Elwood Thompson, was a Navy pilot, and the family typically moved every year or two – Washington, Oregon, California, Texas, Japan, Italy. As a boy, Thompson showed a penchant for things mechanical, especially if there was a bit of associated adventure. In Kingsville, Texas, surrounded by oil fields whose workers communicated with short-wave radios, he hung around the local radio shop and puttered with ham radios. When the repair crews from the shop went out into the oil fields, he tagged along, climbing the rigs to retrieve the broken radios. The family spent three years in Naples, Italy, leaving after Thompson finished his junior year in high school — "the longest we ever stayed in any one place," he recalled. By then, he had moved onto more elaborate projects like building an electrical robot that could pick things up, or welding pipes and coffee cans together to make rockets, launched into the night sky over Naples. Indeed, flight would be a lifelong passion, just as it was for his father. When he was 48, he would pay $12,000 to fly a MiG-29 fighter jet in Russia, an adventure he described as equal parts G-force and adrenaline, doing rolls, loops, Immelmanns, hammerhead stalls, and inverted stalls. Not known as a public speaker or a writer, Thompson nonetheless felt compelled to write a journal of the experience, which he posted on his Web site, "We took off. Full afterburners. It was like somebody kicking me in the kidneys."

His father was posted in the United States, at a naval base in San Diego, in time for Thompson's senior year in high school. Thompson's father had grown up in Kingfisher, Oklahoma and attended the University of Oklahoma for one year, but he left college for lack of money in the Depression to join the Navy. Lewis Thompson was emphatic that the brainiest and youngest of his three children should get a university education. "It was never a question of whether I was going to college, just which one," Ken Thompson said. The choice came down to Berkeley or the Massachusetts Institute of Technology, and the difference in tuition and travel costs tipped the balance in favor of the state school.

At Berkeley, Thompson the electronics hobbyist naturally gravitated toward a major in electrical engineering and the Berkeley computer center. He was soon working long hours at the computer center — subsistence computing in a way, helping pay for his tuition, food, and lodging. Thompson thought of himself more as a professional computer handler, working for a living, than as a student. He observed that he got his masters degree in electrical engineering "almost by accident, I was just hanging around." His instructors, he said, signed him up and gave him course credits for his computer center work. In 1983, when he and Ritchie received the Turing Award, which has been called the Nobel prize of computer science, Thompson explained, "I am a programmer. On my 1040 form, that is what I put down as my occupation." He has called programming an addiction of sorts, and it was in the Berkeley computer center that he got hooked. Sitting in the Bell Labs offices years later, he described the appeal as having all the craftsman's satisfactions of making things, without the cost and trouble of procuring all the materials. "It's like building something where you don't have to order the cement," Thompson said. "You create a world of your own, your own environment, and you never leave this room."

Nighttime was when Thompson often had unlimited access to the Berkeley computers, allowing him to program and configure a machine to do whatever he wanted. What he mainly wanted to do was play games, and one of his favorites was computer tic-tac-toe in a four-by-four-by-four layout. It required the player – human or computer program – to calculate far more moves and options than the conventional game. It was as if tic-tac-toe boards, each with 16 squares, were set up in a tier of four boards. The player would try to get four squares in a row either on a single board or by using squares on each of the four boards. With its decision-tree and branching construction, the elaborate tic-tac-toe game was the kind of puzzle that appealed to mathematical game theorists.

Thompson had become acquainted with a leading game theorist in the Berkeley math department, Elwin Berlekamp. The math don and the computer center hacker had a running challenge that went on for months in multidimensional tic-tac-toe, with games played in the predawn hours. Thompson would laboriously program his machine, and then he would summon Berlekamp from bed at 2 A.M. or so. "He'd come down, and it would be me and my computer versus him and his brain," Thompson recalled. Berlekamp, a day person, would show up bleary-eyed and groggy. "But we never beat him," Thompson said. "I'd go off, lick my wounds and program it again. But he'd just come down again, and win."

For Thompson, however, those nocturnal tic-tac-toe games led to a career at Bell Labs. He was targeted by the labs' recruiters because of an unsolicited recommendation by Berlekamp. When the company recruiter showed up at Thompson's Berkeley apartment, Thompson said he would be glad to take the free trip east – he had friends he wanted to visit – but he was not really interested in working for the telephone monopoly's research lab. When he showed up at Bell Labs, Thompson underwent a two-day regimen of interviews with representatives from several departments. Most of the research departments, he said, had the same attitude toward computers that much of corporate America had toward the Internet in the 1990s – they knew they had to embrace the new technology, but they were basically clueless. The Bell Labs' "computing sciences center," he recalled, was the great exception. It seemed more open, informal, and populated by researchers eager to explore the frontier of the new science of computing wherever it might lead. "It was just a different place," Thompson remembered. "I thought they were cool people."

Dennis Ritchie took a very different path to computing and to Bell Labs. Ritchie was reared in the comfortable suburbs of New York, and his father, Alistair E. Ritchie, was a Bell Labs scientist. As a freshman at Harvard, he attended a lecture by Jean Sammet, who described computing on the UNIVAC. His interest piqued, Ritchie later took Engineering Sciences 110, a course that was an introduction to computing. He visited the local IBM office, where he was given a FORTRAN manual. His early programming was mostly applied math – programs, for example, that might employ differential equations to calculate the flow rates of fluids through a cylinder. But while Ritchie came to computing by way of math, instead of electrical engineering, he found he was more interested in real computers than in the theoretical ones of math models. He continued to study toward his Ph.D. in mathematics, yet he never got his doctorate, even though he finished his dissertation and defended it. But afterwards, he was too disinterested to do the necessary paperwork to obtain the degree. He remembered the episode more than three decades later as "one of those things that one regrets doing, or not, in this case, in relative youth." At the time, however, he was too interested in moving on to a career in computing to care.

While at Harvard, Ritchie landed a part-time job at MIT, working on the university's time-sharing experiments. During a couple of interviews at Bell Labs, Ritchie – slender and bearded – recalled those days fondly. He noted that the early Compatible Time-Sharing System, though primitive in many ways, did allow a user to have an open connection to anyone else who was also on a terminal – the equivalent of today's "instant messaging" so popular on the Internet and on on-line services. Ritchie then worked on MIT's Project MAC, a larger time-sharing system funded by the Pentagon's Advanced Research Projects Agency (ARPA). CTSS and Project MAC were experiments to develop the technology for democratizing computing and to study human-computer interaction. The computing resources were shared in a tiny circle by modern standards. The first Project MAC system, for example, could handle 30 users sitting at typewriter terminals at any one time. Yet Project MAC, especially its next-generation time-sharing system, Multics, would have a lasting effect on the people who worked on it, and the course of computing itself.

When he left Harvard, Ritchie was recruited by the government's Sandia

National Laboratories, which conducted weapons research and testing. "But it was nearly 1968," Ritchie said, "and somehow making A-bombs for the government didn't seem in tune with the times." Instead, Ritchie joined Bell Labs in 1967, a year after Thompson. Soon, both of them were mired in the Multics project. The CTSS and initial Project MAC time-sharing systems ran on IBM mainframes, and both MIT and Bell Labs had close ties to IBM. Yet MIT and Bell Labs were also committed to time-sharing, while IBM seemed to regard it as an academic curiosity instead an important new direction in computing. The IBM mainframes were designed for the long, continuous runs of batch processing, instead of the dynamic interactive model of time-sharing. In 1964, Project MAC informed IBM that its new 360 was ill-suited for time-sharing, and it chose a General Electric machine instead. Soon afterward, Bell Labs opted for a General Electric computer for its time-sharing needs.

By 1965, Bell Labs had joined GE and MIT on the Multics project. The Multiplexed Information and Computing Service was the most ambitious time-sharing effort of its day, designed for 1,000 connected terminals and 300 users at any one time. GE supplied the hardware, while Bell Labs and MIT worked on the system's software. AT&T could not get in the computer business because of a 1956 antitrust consent decree, in which it agreed to limit itself to "common carrier communications services." But Ma Bell understood that computer technology was central to its future. The phone company's electronic switching systems, even then, were essentially large computers. Today, some of the most complex, sophisticated software in use is found in digital switches, where it acts as a kind of traffic cop to direct an endless rush of voice and data messages. While AT&T would be kept out of the business until it agreed to be split into separate companies in 1984, the company was free to develop computer technology for its own needs and engage in research projects like Multics. And time-sharing looked like just another network to the Bell system.

Multics was an effort to deliver large-scale time-sharing efficiently. An ambitious and worthy goal, but Thompson said Multics suffered from a bad case of "second-system syndrome," adding too many new things at once to the time-sharing idea. Trying to get three geographically dispersed and culturally disparate institutions to cooperate certainly did not help. In 1969, deciding that Multics was a costly mistake, Bell Labs dropped out of the project. Yet the Bell Labs team had become attached to the MIT-style of time-shared computing. In those systems, from CTSS to Multics, Ritchie explained, the users shared a connection to the computer and its storehouse of information and programs. This simultaneous use of a shared resource, he said, meant "cooperation and collaboration is much easier. The stuff you're working on is out there for all to see."

Indeed, Unix was created to recapture the collaborative computing environment of Multics, though simply and inexpensively. Like Multics, Unix was designed for interactive systems based on the notion of terminals connected to a central computer. The tree-like structure of Unix, its command names and "command interpreter," or shell, are similar. "The common view is that Unix was made in reaction to Multics, but there is an enormous amount of cultural parentage," observed Ritchie.

Even the name of the operating system – Unics, originally but the "x" was soon added, the first time the name was written as a label on a paper tape – was a "very weak pun" on Multics, according to Brian Kernighan, the Bell Labs researcher who apparently coined the term. "The idea was that Unix was one of what Multics was many of," he said. An early member of the Unix group, Kernighan, like Ritchie, had worked on Project MAC. "Unix would not have happened without that lineage back to Project MAC, without people getting hooked on that kind of collaborative computing."

The story of Unix shows, once again, the yin–yang relationship between software and hardware, so different yet so interdependent. Software breakthroughs have generally come at the stress points – when hardware advances open the door to new possibilities, or when a leap of innovation is needed to overcome a frustrating hardware hurdle. Unix would be an example of the latter.

After the Multics project was abandoned, Thompson and Ritchie lobbied their superiors to let them build their own time-sharing system and to purchase a new Digital PDP-10 machine for the purpose. The Bell Labs management turned down the proposal, but the Multics refugees did not give up. There was a make-do alternative at hand, which they had located thanks to Thompson's penchant for games. The pace of work on the Multics project had afforded Thompson the spare time to develop a computer game called "Space Travel." The game simulated the motion of the planets and moons of the solar system. By typing commands, a player could cruise through space, and even land on one of the planets. It was a hit with the computer science researchers, but the Bell Labs management was not cheered to learn that costly time on its big GE 635 computer was being consumed by a space game. So Thompson found a Digital PDP-7, an outdated minicomputer but with an excellent display terminal for the computer-screen space excursions. He and Ritchie rewrote Space Travel for the PDP-7 – good practice for grappling with an under-powered machine.

For some time, Thompson, Ritchie, and a few other Bell Labs programmers had been thinking about how to design a streamlined file system for time-
sharing. Ritchie called it the "chalk file system," having resulted from a series
of marathon chalkboard discussions at the labs. In the summer of 1969, Thompson began programming the first version of Unix on the PDP-7. His wife Bon-
nie had taken their baby son out to see Thompson's family in California for a
month. While she was away, Thompson gave himself one week for each of the
four main components of the first-generation Unix system: the operating system kernel, the shell, the editor, and the assembler for translating the program
into machine language. This was the birth of Unix, but Thompson regards the
feat mainly as implementing a bare-bones version of a time-sharing operating
system he long pondered. Unix started life humbly indeed, as the programming
glue for a two-station time-sharing system. "It began on this incredibly feeble
piece of hardware," Ritchie noted.

The hardware corset helped ensure that Unix was elegant, compact, and simple. Not getting the larger PDP-10 when they asked was probably a blessing in
disguise for Thompson, Ritchie, and Unix. "I think it was a very good thing
indeed," said Kernighan, who joined Bell Labs in 1969 and was one of the ear-
liest users of Unix.

When Unix was first presented to the outside world at a conference in October 1973, Ritchie and Thompson stated, "Perhaps the most important achievement of Unix is to demonstrate that a powerful operating system for interactive use need not be expensive either in equipment or in human effort: it can
run on hardware costing as little as $40,000, and less than two man-years were
spent on the main system software."

The lean efficiency of Unix reflects the process of its creation – a tiny team
over the years, working at its own pace, unguided by market research or thought
of customer requirements. "We didn't have that mentality," Thompson
observed. "We did it for ourselves. We were arrogant czars in that sense."

"Unix wasn't really a goal in itself," he added. "It wasn't like the Manhattan
Project, where you explode something and that is your goal. If there was a goal
with Unix at the start, it was as a tool, to build support for our research, the
other things we wanted to do."

So Unix was the product of a gradual process of technical gestation. That
allowed Unix to evolve according to a timetable removed from the commercial pressures. No features were mandated, but radical overhauls were also pos-

sible because there was no audience of customers to upset. Good ideas were folded into the operating system as they came to fruition in a research setting.

The big breakthrough was implementing the concept of software "pipes" in Unix, which occurred in 1972. Doug McIlroy, a Bell Labs scientist and manager, had long been pushing the idea of connecting programs, as if in streams of data. For years, Bell researchers had been experimenting with software that would help overcome the restrictions of traditional batch processing. In the early 1960s, for example, Bell programmers created Snobol, a language for manipulating data strings – a list-processing cousin of John McCarthy's Lisp language. In an internal memo in October 1964, just as Bell Labs was about to join the Multics project, McIlroy wrote, "We should have some ways of connecting programs like garden hose – screw in another segment when it becomes necessary to massage data in another way."

In 1972, McIlroy was prodding Thompson to put stream processing into Unix. McIlroy, Thompson recalled, had sketched out some diagrams for his idea – drawings that depicted a complex mesh in which the output of any program could be fed into any other program, any number of times. Each garden hose, then, might have many connectors, as if a tiny software version of the Christmas-tree valves in oil refineries. It struck Thompson as Rube Goldberg-like, but he was impressed. He was "amazed by the premeditation" of McIlroy's design, he said, because it was so different from his own approach to software. "I do things by successive approximation. That's how I program. I just push a program into the program I want it to be."

To Thompson, his supervisor's design was a theoretical fantasy, which totally underestimated the unpredictable complexity of software. "My response was, 'These are programs! You don't even know that they run, let alone how they behave'. . . . I thought it was pie in the sky. In my mind, I had already proven him wrong."

Yet if McIlroy's method seemed misguided, his concept of data streams intrigued Thompson. The boss kept prodding, and the programmer kept thinking. "One night – I don't know why – I got the idea to make it one-dimensional," Thompson recalled. That is, instead of allowing many inputs from other programs into a program in a fairly arbitrary way, Thompson decided to make it one to one. The output of one program could become the input to another program – a linear assembly of fitted "garden hose" rather than mounds of hose. He took the complex general case of McIlroy's design and pared it down to the simplified special case. In programming terms, there would be no loops, no feedback, affording a certain elegant predictability. "If you give it no options, it works," Thompson said. Looking back, Kernighan, who is now a professor at Princeton University, marveled, "What a burst of innovation that was. It just opened all sorts of doors."

The work of implementing pipes into Unix was accomplished in three nights – Thompson alone on night one, joined by Ritchie the next two. Thompson first gave the shell of the operating system a tool for creating a channel that would enable one program to communicate directly with another. That removed the middlemen, or in-between steps, required in running software on an operating system. In the old way, programs were stand-alone entities that would run and then were shunted off to some temporary file – a software holding pen, awaiting new orders to go elsewhere. In the pipe paradigm, things became much more fluid and flexible, with one program accepting a stream of data without hesitation from another. The next two nights Thompson and Ritchie overhauled Unix by changing its internal rules, notations, and tools to exploit the pipe concept.

They worked hunched over the keyboard in front of a PDP-11 machine in a fluorescent-lit low-ceiling room on the sixth floor of Bell Labs in suburban New Jersey – Room 2C644, as if a memory address in some antiquated computer – or from home, since both had remote terminals. Their days typically began late – Thompson often arriving a minute or two before the Bell Labs lunchroom closed at 1:15 P.M. – and they would work until 3 or 4 A.M. Once Thompson solved the riddle of the pipes, the work went quickly, tweaking one Unix feature after another to take advantage of piping. It was an exhilarating sprint. "Those were probably three of the greatest nights of my life," Thompson observed.

The work bore the personal imprint of Thompson the programmer, known for his speed, economy, and vision. "Ken's code," McIlroy said, "is lovely to read, utterly clean, obviously the product of insight, not of debugging." When an idea for an improvement strikes, Thompson tosses out the old code and rewrites from scratch rather than trying to cobble in the improvement. He tends to shun the use of programming shortcuts, known as macros, and graphical aids to code writing. In 1975, a British programmer George Coulouris developed a visual alternative to Thompson's code-editing tool for Unix, which allowed a programmer to view the code on the screen of a video display unit. At Bell Labs, Thompson programmed on a Teletype terminal with a keyboard but no screen, working from a printout and holding the mental image of changes he made in his head. When Thompson came to London, Coulouris showed him his screen editor and he recalled Thompson's reply as, "Yeah, I've seen editors like that, but I don't feel a need for them. I don't want to see the state of the file when I'm editing." Afterward, Coulouris decided to call his visual editing program "em," short for "editor for mortals."

Thompson's programs are accompanied by few comments; the code speaks for itself, concisely. Ritchie speaks of Thompson's "amazing inventiveness in coming up with the right algorithm" for the task at hand. Thompson holds an early software patent, submitted in 1967 and issued in 1971, for an innovative text-matching algorithm. In fifteen pages of description and programming instructions, the patent describes a method for rapidly identifying patterns of text by computer. By employing the algorithm, a program can conduct several search functions in parallel rather than the slower method of one-at-a-time sequential steps. Thompson's bit of programming sleight of hand to identify and compile data very rapidly was a forerunner of the on-the-fly compilation techniques used to make Internet programs run quickly.

Thompson also applied his algorithmic inventiveness for racing through the logic trees in mathematical games like "moo" and "chomp," and later to chess. In 1980, Thompson wrote the software for a chess-playing computer. He and a Bell Labs colleague, Joe Condon, turned out the machine, Belle, in months, and to do that required Thompson to perform magic in an array of special-purpose computer languages but also to work on the wiring. "Ken goes down to the bare hardware," Ritchie said. "He can actually build things." Belle was the first computer to attain the level of a chess master, setting the pattern for later machines like IBM's Deep Blue, which defeated the reigning world chess champion, Gary Kasparov, in 1997.

His boss McIlroy pointed to that first night in 1972, when Thompson began putting pipes into Unix, as a dramatic demonstration of Thompson's "amazing ability to see a project whole." He not only implemented the pipes, McIlroy recalled, but also altered the operating-system shell to exploit them, and then began fixing a handful of Unix utilities as well. Those tasks, thought out piece-meal, McIlroy explained, would have taken other programmers far longer. "A prudent project manager," he said, "would have allocated a few weeks to the job."

The pipes, most of all, made Unix more than just another operating system. The Unix environment came to represent a new way of thinking about programming, called the "tools philosophy." With its pipe framework, Unix encouraged programmers to think of a program as a flexible tool that can be combined with other tools to do a job or build a software application. It is E. F. Schumacher's small-is-beautiful thinking brought to software. It stands in striking contrast to viewing a program as an all-in-one package of related functions, like a 360-style mainframe operating system or, later, Microsoft's Windows operating system or Microsoft's massive applications programs, the Word writing program and the Excel spreadsheet.

Programmers who embraced the tools approach underwent a conversion of sorts. Brian Kernighan, one of the first, described the epiphany: "Hey, I'm not writing programs. I'm making tools others can use and build on." The Turing Award citation for Thompson and Ritchie observed that the "genius of the Unix system" is that it "enables programmers to stand on the work of others."

At his Princeton office, Kernighan used a simple example to show how the programming pipes in Unix can be used to link up different software tools. Let's say, Kernighan began, that he wants to find out how many people are using the same networked computer he is tapped into at the university. He types in the Unix command, "who," and on his screen pops a long list of work stations attached to the central computer, each on a line which includes the name of the person assigned to each machine. In the old days at Bell Labs, he says, the "who" tool was a quick way to find who was around. But the Princeton who-list is hundreds of names. In order to find out how many people are on the network, Kernighan pipes the results of the "who" program into a word-counting program, which actually counts three things: lines, words, and characters. The programming instructions are: who | wc. (The single vertical line is the symbol for a pipe.) The word-counting program counts 428 lines, so there are 428 connecting to the same central computer at Princeton. Next, Kernighan says, he may want to find out if a friend – say, Joe Smith – is one or more of those connections (a single person can have several connections). Well, there is a Unix text-pattern searching tool called grep. To find out if Joe is currently logged in, use a pipe: who | grep joesmith. To find out how many times Joe Smith has logged onto the hub computer that day, roll out another pipe: who | grep joesmith | wc.

The pipes enable both flexibility and programming speed. In 1973, in a classic display of the tools approach, Steve Johnson, a Bell Labs researcher, threw together a spell-checker program with about a half-dozen Unix tools including some sorting and conversion utilities, and a dictionary-style word list. It took him about a minute. "Steve didn't have to write a line of code except the pipeline," Kernighan recalled. The Unix environment gives programmers more freedom to quickly add, subtract, or combine tools. It becomes easier to experiment with different combinations of software. In the Unix world, for example, the spell-checker and word-counting functions in a word processor are separate programs, so there are alternatives to choose from. In Microsoft's Word, however, there is one spell checking tool and one word-counting tool – the ones Microsoft chooses to include.

The two approaches, poles apart, are sometimes defined as if rival political camps, if not religious movements. To a certain breed of programmer, the Unix tools philosophy is seen as both practically and morally superior, with its promise of freedom to experiment with unlimited choice. It is ideal for the kind of programmer who savors a free-form, unstructured environment and who, as Thompson described his own style, works by successive approximation, pushing the program little by little into a finished design. Thompson, not surprisingly, believes in the culture he helped create. Of the Microsoft approach, he said, "I just think the philosophy is bankrupt. The attitude toward users is demeaning. It treats users as dummies, so it makes all kinds of things not programmable."

Yet to others, the Unix tools system is a formula for confusion and unnecessary work. These programmers and users prefer the convenience of having Microsoft supply the basic software, so they can build their application on top of it or just do their work. They don't want to dig down into the software on their own. Even Ritchie concedes that the tools approach is not for everyone. "The idea behind the tools approach is that the pieces are supposed to be simple enough so they are understood," he explained. "But we may have not succeeded in making things as simple as we had hoped. Not everyone is equipped to do this kind of thing. It may be a model that goes only so far."

Unix has certainly gone far further than anyone could have possibly imagined back in the early 1970s, when the fledgling operating system was all but unknown outside Bell Labs. Its first practical use was as a word-processing system to prepare patent applications for Bell Labs – to lighten the workload on three full-time typists. Unix spread first to university researchers who heard about the lean, efficient operating system that seemed ideal for collaborative work on minicomputers. As a regulated monopoly, Ma Bell was forced to license its technology – a policy that ensured innovations like the transistor and laser, not just Unix, were widely dispersed. Ritchie, Thompson, and others lobbied to have the Unix licensing terms to universities be extremely inexpensive and open; AT&T's lawyers agreed, seeing little harm from being so permissive, and little to be gained from the software research project, other than goodwill at universities.

University researchers loved Unix, and it flourished at leading computer science departments in the United States like Berkeley, Stanford, and Carnegie Mellon, and gained a foothold abroad as well, mainly in Britain. Unix, typically running on PDP-11 minicomputers, drastically reduced the cost of computing, not only in money but in programmers' time. Designing and debugging a program on a Unix system was perhaps ten times faster than working on a batch-processed mainframe, estimates George Coulouris, a professor emeritus at the University of London's Queen Mary and Westfield College, where the first Unix system in Europe was installed in late 1973.

The Bell Labs licenses laid bare the intellectual content of Unix by distributing the operating system's "source code" – the programming instructions humans can read, as opposed to a program's "binaries," the 1's and O's that the machine reads. The Unix manuals distributed with the code were models of clarity and precision. They were written for experts, often requiring several readings for even skilled programmers to understand, but they were comprehensive and fully documented every aspect of the operating system. The Unix manuals even included a section on "bugs" in the software, a level of candor and honesty that would have been unthinkable coming from commercial computer companies. In the 1980s, things would change. AT&T tried to make Unix a conventional software product, and rivals jumped in with their own versions of Unix, setting off a marketing and technical tussle that came to be called the "Unix wars." But during the 1970s, Unix was a different culture. To young university computer scientists, chafing at the restrictions of traditional computing, the Unix environment was exhilarating, almost addictive. "It was like handing out heroin on the playground," Kernighan said.

The C programming language was made for Unix. It sprang from the same place, and it shares a similar mentality. And, in fact, it was the creation of the same partnership. "Unix is Ken Thompson with an assist from Dennis Ritchie. And C is Dennis with an assist from Ken," observed Kernighan, who was an important contributor to Unix and coauthor with Ritchie of *The C Programming Language*.

C traces its origins back to MIT and the Project MAC. Its "grandfather" was BCPL, designed by Martin Richards, a British computer scientist who was at MIT briefly during the 1960s. BCPL was a little language developed as an offshoot of the Multics project; the main Multics language, PL/I, was an overstuffed attempt to combine elements of FORTRAN, COBOL and Algol. When he created the first version of Unix, Thompson wrote it in an assembly language tailored to the small minicomputer he was working on. He programmed in assembly language because memory was so scarce on the feeble machine first used to run Unix. Afterwards, though, Thompson decided that Unix should have a programming language of its own to make it easier to write programs, but the language had to work on an undernourished piece of hardware. His solution was to design a stripped-down version of BCPL, called B, which Ritchie later described as BCPL squeezed to run on a computer with a meager ration of memory and "filtered through Thompson's brain."

Both BCPL and B were what is known as "typeless" languages, meaning that the language does not distinguish between different types of data. In older computers, pieces or blocks of data – often called "words" – were all the same size. Smaller machines, for example, might handle only 16-bit words, while larger machines handled 36-bit words. But when the Digital PDP-11 minicomputer was introduced in 1970, the machine was designed to recognize and process more than one size of data objects. That opened the way to a language which specified different types of data – integers, floating point numbers, alphanumeric characters, and dates, for example. With C, Ritchie fashioned in 1972 a language to make that capability of the machine visible and usable to programmers.

C comes from the same culture as Unix – an elegant tool, but one designed with the skilled professional in mind. It is a language that resides fairly "close to the machine" – its abstractions not too far removed from the understanding and operations of the computer. Still, C is a higher-level language in that it is not wedded to a specific machine. Yet it was designed to allow programmers to do heavy-duty "systems" programming, working on such basic software plumbing as operating systems, compilers, and the like. Before C, systems programming was not really done in higher-level languages. C did for systems programming what FORTRAN did for scientific and engineering computation – it made it easier to communicate with the machine, lowering a barrier to computing in that discipline. "It hit the right ecological niche, the right tool at the right time," Ritchie explained.

Ritchie and others labored to broaden its ecological niche by making C run on different machines. In that way, C also spread Unix from minicomputers to work stations to supercomputers.

C places few restrictions on the skilled programmer's freedom, and few safeguards to keep the less-skilled out of trouble. C allows the programmer, using a "pointer," to dip into the machine's memory and fiddle with the data there in any way the programmer chooses – appropriate or not – performing some trick of software wizardry or crashing the machine. "Pointers do have their dangerous aspect. They are fairly easy to misuse," Ritchie observed cheerfully, suggesting that C, like Unix, is not meant for everyone. For a decade, C was closely tied to Unix. C spread with Unix as the operating system became more widely used, but C also help spread Unix by making it more accessible, easier to program for. During the 1980's, C jumped across to the personal computer programming community, becoming a mainstream language in the PC industry as well. "Sometimes, the tools that you design for yourself scratch the itch of a lot of other people," Ritchie noted.

Both Unix and C obviously filled needs at a certain time in the evolution of computing – the right "ecological niche," in Ritchie's phrase. Timing, luck, and happenstance all played a role, but they were also technologies that took root, flourished, and endured because they were a deft combination of innovation, taste, and utility. They were also the result of a rare partnership between two people of different backgrounds, temperaments, and abilities. Ken Thompson was more the pure programmer – "light years ahead of the rest of us," Kernighan recalled. Dennis Ritchie was also a brilliant computer scientist, but of a different type – steadier, more methodical, and systematic. Ritchie was the writer and speaker of the two. He had the patience to navigate the politics required to make a language like C an internationally ratified technology standard. Doug McIlroy, who supervised the pair for years, observed that Ritchie's talents, while considerable, were "solid" and "intelligible." Thompson's ability, he said, was of a different kind, "undeniably genius class," which delivered flashes of "unexpected and exuberant creativity."

Whatever the precise chemistry of the partnership, the intellects of the two often worked as one. In 1983, when accepting his Turing Award, Thompson called the collaboration with Ritchie "a thing of beauty" and then gave a little example to show how similarly their minds worked in code. Only once in a decade of working together, Thompson recalled, could he recall a communication breakdown between the two. "On that occasion," Thompson said, "I discovered that we had both written the same 20-line assembly language program. I compared the sources and was astounded to find that they matched character-for-character."

"The result of our work together," Thompson concluded, "has been far greater than the work we each contributed." Fitting, it seems, that the Unix tools philosophy – a system for enabling programmers to stand on the shoulders of others – so accurately describes how its creators worked.

5

Programming for the Millions: The BASIC Story from Dartmouth to Visual Basic

IN 1957, THOMAS KURTZ WAS A YOUNG PROFESSOR at Dartmouth College in New Hampshire, but also a frequent commuter to the Massachusetts Institute of Technology, where he could get his hands on a big IBM computer. His first taste of computing had come six years earlier as a graduate student, when he got a summer job at a government research center in Los Angeles. That early experience, Kurtz recalled, caused him to drift toward computing, even though he went on to get his Ph.D. in statistics from Princeton University and Dartmouth had hired him to teach statistics. In those days, computing was a new, uncharted field whose potential significance – economically, socially and culturally – was just beginning to be appreciated. But Kurtz had a simpler motivation as well. "For people who program, programming is *fun*," he explained more than four decades later, speaking of the rigorous, intellectually–consuming pleasure of talking to the machine – often to the "consternation of those around you who might want a minute of your attention."

The early programming Kurtz did was in assembly language, a separate one for each different machine. At the MIT center the computer was an IBM 704, and he learned its assembly language, SAP, for Share Assembly Language. In 1957, FORTRAN arrived, though there was a prejudice against the so-called higher-level language at first. FORTRAN, many programmers said, was for the lesser practitioners of their craft. Real programmers wrote programs in assembly and, they believed, surely saved expensive machine time by doing so. So, when Kurtz had a project that involved programming a vast array of statistical calculations, he proceeded in the SAP assembly language. After months of trying, he admitted failure, having consumed "one hour of valuable 704 time and untold hours of my less valuable time." Abandoning assembly language, he then tried the disparaged, inefficient FORTRAN. "The answers appeared," he recalled. "About five minutes of computer time were used. This lesson – that programming in higher-level languages could save computer time as well as person time – impressed me deeply."

Kurtz and his Dartmouth colleague and mentor, John Kemeny, went on to devise an easy-to-learn, higher-level language called BASIC, for Beginner's All-purpose Symbolic Instruction Code. The language was created mainly to give liberal arts college students a feel for computing, and it was first used in 1964 on the Dartmouth time-sharing system. BASIC became the leading programming language on time-shared computer systems of the late 1960s and early 1970s, used by college, high school, and even elementary school students. For the young computing upstarts who created the personal computer industry, BASIC was their first programming language, and it shaped their approach to computing. The personal computer newcomers grabbed BASIC and shrank it, modified it, and extended it. They moved quickly, before any standards group set official rules for BASIC. As a language, BASIC seemed ideal for the wide-open, exuberant, entrepreneurial days when the PC industry was aborning. BASIC was simple, flexible, and malleable. Its few informal rules were broken with impunity. "BASIC was an open city, Shanghai a hundred years ago," observed Alan Cooper, an early PC programmer. "There were no laws."

By the late 1970s, there were dozens of different versions of BASIC out in the marketplace. But thanks to clever programming and shrewd business tactics, one came to dominate, Microsoft BASIC, which was the software giant's first product in 1975. One of the axioms of the computer business is that there is not much money to be made in programming languages, which are merely tools, after all. But Microsoft's strategy over the years has revolved around establishing an industry standard technology – or a technology "platform" – in large part by enticing software developers to write programs that run on top of Microsoft's platform. The pattern started with Microsoft BASIC and was extended with its Windows operating system. And the success of Windows owes a large debt to a home-grown variant of BASIC called Visual Basic, a programming tool of the 1990s. The first version of Windows was shipped in 1985, but the graphical operating system did not become a big winner until 1991. Microsoft had improved Windows, but 1991 was also the year the company brought out Visual Basic, which made it far easier to write programs that ran on Windows. "Over the years," observed Bill Gates, "BASIC in all its forms has been the key to much of our success."

The original motivation behind BASIC – and the Dartmouth time-sharing system – was both educational and cultural. In the early 1960s, it was becoming clear that computers would be more than just big calculators, but also data and symbol processors that would affect business and government as well as science. Kemeny and Kurtz were influenced by the concerns raised in C. P. Snow's Cambridge University lecture in 1959, "The Two Cultures," which dissected the difference between scientific and literary intellectuals, and noted the danger of the schism. "Those who know how the system worked," Kurtz observed, "and those who merely responded to it." He and Kemeny thought something should be done at Dartmouth, an elite Ivy League school. Only a quarter of the Dartmouth students majored in science or engineering, the group mostly likely to be interested in computing. But "most of the decision makers of business and government" typically come from the less technically inclined 75 percent of the student population, Kurtz wrote in a paper in 1978. "We wondered, 'How can sensible decisions about computing and its use be made by people who are essentially ignorant of it?' This question begged the conclusion that nonscience students should be taught computing."

Dartmouth was a very early convert to the notion that computing should be part of the liberal arts curriculum, but other educators, often in the Ivy League, have advocated the same stance. In 1979, for example, Joseph Traub departed as the dean of Carnegie Mellon University's computer science department, one of the leading programs in the nation, and went to Columbia University to build its computer science department. At Columbia, Traub recalled that his challenge was to "convince one of the great arts and science universities in the United States that computer science was really central," which he did. And today, Brian Kernighan, a member of the original Unix team at Bell Labs, teaches computing to freshmen at Princeton University, and he echoes the sentiments heard at Dartmouth in the 1960s. "The people making policy decisions in our country ought to understand computing because it is a pervasive part of our lives," Kernighan said. "And until you do battle with the machine, you don't really understand how precisely you must talk to a computer to make it do what you want. And you don't understand all the things that can go wrong." So in his class, Kernighan added chuckling, "I make the little varmints program" – in Visual Basic, incidentally.

In 1962, Kurtz approached Kemeny, the chairman of the math department, with what Kemeny called the "outrageous suggestion" that every Dartmouth student should learn to use a computer. To Kemeny, the outrageous aspect of the suggestion was that it seemed so impractical at a time when computing was typically done by batch-processing punched cards. But Kemeny agreed with Kurtz about the importance of computing. John Kemeny was both a mathematician and a philosopher, who later became the president of Dartmouth. His family had emigrated from Budapest to New York City, and after high school Kemeny entered Princeton and majored in math. In 1943, while a junior in college, Kemeny was drafted and dispatched to Los Alamos to work on the Manhattan Project. He was one of the laborers in the project's "computing center," where the calculations necessary for designing the atomic bomb were churned out on desktop calculators by crews working three eight-hour shifts, 24 hours a day.

Kemeny, who died in 1992, would joke that a year's worth of work by his Los Alamos number-crunching team of 20 people could be done in a single afternoon by a Dartmouth sophomore on the school's time-sharing system of the 1960s. After the war, Kemeny returned to Princeton, earned his doctorate and became Albert Einstein's research assistant at the Institute for Advanced Study. Einstein, Kemeny once recalled, used to say that the real value of computers would be as they moved from being big numerical calculators to being symbolic processors. The notion made an impression on Kemeny. And it was the software, as well as greater processing speed, that enabled machines to move from manipulating arithmetic symbols to manipulating symbols of other kinds – symbols for words, physical phenomena, even ideas. That is why computers could be programmed to play chess, simulate weather patterns and, eventually, unlock the secrets of the human genetic code. Even BASIC's name nodded toward that further horizon of computing, with the "S" standing for symbolic.

Yet how to deliver the computing experience widely to university students remained a puzzle. At the time, the MIT time-shared computing experiment was underway, and John McCarthy was already an enthusiastic advocate. While visiting MIT, Kurtz talked to McCarthy, who counseled without hesitation, "You guys ought to do time-sharing." Upon reflection, Kurtz and Kemeny agreed, and with the support of the Dartmouth administration and funding from the National Science Foundation, they began designing a time-sharing system of their own.

A programming tool for computer novices did not really exist in the early 1960s. The higher-level languages available, such as FORTRAN, COBOL and Algol, were a big step up from assembly languages in simplification, but they were designed mainly for specialists in certain fields like science, engineering, or business. They included grammatical rules that were not intuitively obvious to liberal arts students. Before BASIC, Kurtz and Kemeny had experimented with a few teaching languages. Kurtz and two students designed a language called Scalp, for Self Contained Algol Processor. In 1962, Kemeny and a student designed another teaching language called DOPE, for Dartmouth Oversimplified Programming Experiment. Both Scalp and DOPE contributed ideas about how to simplify and streamline programming that would find their way into BASIC.

From their own computing experience and watching their students, Kurtz and Kemeny agreed on eight design principles for BASIC. The new language, they decided, should be easy to learn for a beginner, and general purpose so that all kinds of programs could be written in it. Among the other requirements were that the language should give easily understood error messages, and the language would permit extensions and advanced features, but those should not be added in a way that complicated things for the novice user. No understanding of the hardware, they added, should be necessary.

To simplify presentation, BASIC had one instruction per line. Each line began with a statement of operation, or command. There were 14 BASIC commands including LET, READ, DATA, INPUT, GOTO, and IF THEN. To simplify things further, Kurtz and Kemeny included default formats for printing and for the number of decimal places to carry out a calculation, so that a user did not have to program those details. Each line of BASIC code was numbered. The line numbers enabled a student to correct an error or alter a program by retyping a line using the same line number, or erase a line by typing its number and leaving it blank. Below is an example from the Kemeny and Kurtz text, *BASIC Programming,* a simple arithmetic problem, dividing 147 by 69.

```
DIVIDE
100 READ N, D
200 DATA 147, 69
250
300 LET Q = N/D
400 PRINT N, D, Q
450
500 END
RUN
```

In this simple case, the BASIC program – DIVIDE – tells the computer to read the numerator ("N") and denominator ("D") for which the respective data are 147 and 69. The calculation formula – the LET line of instruction – tells the computer to let the quotient equal the numerator over the denominator. The computer is then told to PRINT out the results, and informed that is the END of this particular chore. The blank lines are ignored by BASIC. The authors chose to use blank lines in lines 250 and 450 merely for aesthetic purposes, to make the program easier to read. To get the computer to execute the program, the user types the RUN command. And at Dartmouth, Kurtz noted later, the programming was always done "on slow Teletypes with yellow rolled paper."

```
The computer replies:
DIVIDE 02 MAY 79 17:53
147 69        2.13043
0.114 CRU 0.055 SECS
READY
```

The computer responds by printing out, on the yellow paper roll, the requested information – the numerator, denominator, and calculated quotient, which in this case is 2.13043. It gives the date and time, May 2, 1979, at almost 6 P.M., when the program was executed for the 1980 Kemeny–Kurtz text. This particular time-shared computer then tells the user the number of "computer resource units" (CRU) the program took, and the amount of computer time consumed in seconds. When the computer is ready to receive another command, it types READY.

The Dartmouth time-sharing system first stirred to life on May 1, 1964, executing a simple program in BASIC at 4 A.M., bowing to the nocturnal tradition of computing. By the fall of 1964, Kurtz and Kemeny began their freshman programming course. In the first semester, the course consisted of three lectures introducing the students to BASIC, and each student then got 30 minutes of reserved time on a teletype machine connected to the time-sharing system. Each student was required to program and debug four problems. In the second semester, Kurtz and Kemeny modified the curriculum a bit. Each student was given 45 minutes of teletype time a week. "We had underestimated how badly our students type," Kemeny and Kurtz wrote in a 1968 article in *Science* magazine. They also cut back the lectures. "We have found that two one-hour lectures are entirely adequate to introduce the novice to BASIC. By the end of the second hour he is raring to write his first program."

Clearly not wanting to check the thousands of student programs written each year, Kurtz and Kemeny wrote a program to handle that chore. When a student typed TEST, the software took control of the student's program, and either gave its approval or offered clues about how to debug the program. The approach, Kemeny and Kurtz wrote in 1968, had the "very great advantage that only the student knows how many stupid mistakes he made before his program was accepted."

The liberal arts students embraced the style of computing offered at Dartmouth. Though not a requirement, the computer programming course was taken by 80 percent of the students. The interactive nature of time-shared computing, as opposed to punched cards and batch processing, was essential to the appeal. By today's standards, the General Electric time-shared computer would be considered lethargic to the point of being defective. "We have found," Kemeny and Kurtz observed, "that any response time that averages more than 10 seconds destroys the illusion of having one's own computer." Yet, sitting at teletype terminals in the 1960s, the Dartmouth students felt that the machine was alive, responding individually to each student's keyboard commands. BASIC was the language in which students of all ages got a sense of "personal computing" in the 1960s and early 1970s.

In the mid-1970s, with the arrival of the microcomputer, the concept of personal computing suddenly changed from the illusion provided by time-sharing to the reality of literally owning one's own machine. Time-shared computers became obsolete, but BASIC made the transition. It became the programming language of choice for the microcomputer revolution, though in a cacophony of dialects at first. Dennis Allison, an instructor at Stanford University, wrote one of the first, called Tiny BASIC. Allison did it at the urging of his friend Bob Albrecht. A computer engineer, Albrecht had fled Control Data Corporation in the 1960s, uneasy with the industry's emphasis on serving institutions and corporations instead of individuals. He came to San Francisco and was soon at the center of the alternative computing culture of northern California. He started a tabloid, *People's Computer Company*, which spread the gospel of computing for the masses. The publication attracted a following, and Albrecht placed it in a nonprofit foundation simply called PCC, which opened a walk-in computer center in 1970 in a shopping center in Menlo Park, a pre-microcomputer creation allowing members of the public to use time-shared computers.

Albrecht found a kindred spirit in Dennis Allison, and the Stanford lecturer was a regular at the nonprofit computer center. It had some keyboard terminals connected to a Digital Equipment PDP-8 minicomputer, and later a few more when Hewlett-Packard donated computing time via phone-line links. The walk-in traffic ranged from elementary school children through adults. The subsidized price of computer time was 25 cents an hour. The computers were used for math puzzles, small-business programs, and all kinds of games, from very simple ones like Stars and Snark to more elaborate text-based adventure or role-playing games like Wompit and Hummarabi. It was all done in BASIC. The PCC storefront computer center was also a clubhouse for an evolving community of computer enthusiasts. Children's birthday parties were held there, and it was the venue for regular pot luck dinners. It was the early home to a side of the computing culture that regarded machines and code as tools of liberation. Its members were typically social liberals who were advocates of free speech and were antiestablishment, anticorporate, and anti-Vietnam War. Ted Nelson, author of *Computer Lib*, frequently dropped in at the PCC pot luck dinners. The clarion call of Nelson's self-published manifesto captured the message and the passion of the computing counterculture: "You can and you must understand computers now!"

Sitting at an outdoor table on the Stanford campus over a boxed sushi lunch, Allison recalled, "The social impact of the PCC was incredible. In many ways, it shaped the socially-conscious wing of the personal computer movement." Times changed, and so did some of the people, but Allison defined the enduring values of that culture as "openness, sharing and not mincing words." Tiny BASIC was created and distributed in that spirit. The MITS Altair micro-computer, which cost $400, had appeared on the cover of the January 1975 issue of *Popular Electronics* magazine. Bob Albrecht saw its potential as an afford-able, general-purpose computer, based on the Intel 8080 microprocessor. Yet it was just dead circuitry, without software to bring it to life to do useful or fun things. Albrecht prodded Allison to develop a shrunken version of BASIC so children might be able to use the Altair and other microcomputers that were just beginning to appear.

Allison went to work, and his early efforts were published in the PCC newsletter and a sister publication, *Dr. Dobb's Journal* (Dobb's was a contraction of Dennis and Bob, the first names of Allison and Albrecht.) "The Tiny BASIC project at PCC represents our attempt to give the hobbyist a more human-ori-ented language or notation with which to encode his programs," Allison wrote. In keeping with the PCC credo, Tiny BASIC was seen as a tool for early com-puter education and tinkering. An article in the PCC newsletter, by Allison "& Others," made its imaginary pitch to the playground:

> Pretend you are seven years old and don't care much about floating-point arithmetic (what's that?), logarithms, sines, matrix inversion, nuclear-reactor calculations, and stuff like that. And your home computer is kind of small, not too much memory. Maybe it's a Mark-8 or an Altair 8800 with less than 4K bytes and a TV Typewriter for input and output.
>
> You would like to use it for homework, math recreations and games like NUMBER, STARS, TRAP, HURKLE, SNARK, BAGELS.
>
> Consider, then, Tiny BASIC.

Serious hobbyists – not seven-year-olds – took the Tiny BASIC that Allison started and modified it and extended it, making it a more useful language. Tiny BASIC was a simple tool that enabled thousands of programmers to begin exper-imenting with microcomputers. And there were no restrictions on programmers taking Tiny BASIC and doing whatever they wanted with it. Allison and Albrecht had published it for others to do with as they wanted. There was no thought of trying to make money from it; they believed in free speech and free software.

Across the country, in Cambridge, Massachusetts, two young entrepre-neurs were taking a very different approach. The often-told story of Paul Allen

and Bill Gates, seeing the January 1975 issue of *Popular Electronics* and then immediately rushing to program a commercial version of BASIC for the MITS Altair,
is the stuff of legend. Yet the "single moment of epiphany" came after years of programming and watching. "The event that started everything for us business-wise
was when I found an article in a 1971 electronics magazine about Intel's 4004
chip, which was the world's first microprocessor," Allen explained years later. "It
made me realize that computing was going to be a lot cheaper than it had ever
been and that a lot more people would have access to computers."

Allen and Gates had attended the private Lakeside School in Seattle, where
computing became their passion. Gates began programming as an eighth-grader,
in BASIC on a time-sharing system. The two had even started a company together,
Traf-O-Data, which built a computer for processing highway traffic-flow information. Allen and Gates had discussed writing a version of BASIC for the earli-
est Intel chips, but decided to wait. The MITS Altair, with its Intel 8080 chip, was
the processing engine they had been waiting for. At the time, Allen had dropped
out of Washington State University and was working in Boston for Honeywell,
while Gates was at Harvard. "Bill and I were anxious to start our own company,"
Allen recalled. "We realized that we had to do it then or we'd forever lose the
opportunity to make it in microcomputer software."

The Altair had been announced, but it was just beginning to become available, and Allen and Gates could not wait. So Allen wrote a program that simulated the Altair's Intel 8080 microprocessor and also most of the development
tools. Then Gates, working on a DEC PDP-10 minicomputer at Harvard's
computing center, wrote most of the code for their original BASIC, assisted by
another Harvard student, Monte Davidoff, who wrote the math-processing
routines. Eventually, questions were raised about Gates's marathon use of the
computer center – especially for bringing in Allen, who was not a Harvard student, to work on a commercial project. Gates could have been given an official reprimand, or expelled, but in the end no action was taken. Years later, Gates
would contribute millions to Harvard – more than making amends, it seems,
for bending the computer center rules.

The BASIC that Gates wrote for the Altair packed a lot into a very small space.
It was fast, and had an impressive set of features for a language that initially ran on
a memory board of just 4K, or 4,000 bytes of data (budget-priced personal
computers in 2001 came with 64 million bytes of memory, or 16,000 times
as much). In high school, Gates had written a BASIC interpreter for a minicomputer. He "made massive mistakes in that program," Gates once recalled, but

the experience prompted him to study other programs for converting, or inter-
preting, software written in the BASIC language into code that can be executed
by a computer. By the time Gates sat down to do Microsoft BASIC in 1975, he
and Allen had scrutinized "dozens and dozens of BASICs." The issue, then, was
not whether Gates could write a version of BASIC, but making a BASIC as fast
and flexible as possible in a paltry 4K of memory. "Actually, making a BASIC run
in that little memory is a real feat," Gates recalled in 2001. "Of all the programming I've done it's the thing I'm most proud of."

To create a more powerful language for a microcomputer, Gates and Allen
ignored the Dartmouth rules of BASIC. To save space, they crammed more than
one programming instruction into a line of code. For more control over the
machine, they included so-called PEEK and POKE commands, which allowed a
programmer to first view, and then to manipulate bytes of data in memory directly
– a violation of the Kemeny–Kurtz principle that their higher-level language
should liberate the programmer from machine details. Microsoft's BASIC for the
Altair was a deftly engineered combination of Dartmouth BASIC and perfor-
mance-enhancing modifications, most of which were originally done elsewhere.
Programmers at Digital Equipment, for example, had first implemented PEEK and
POKE in BASIC on a minicomputer time-sharing system in 1971. But doing
such things in a microcomputer, Gates insists, was a different kind of challenge. At
the time, he noted, the Digital Equipment minicomputers were considered cheap
computers, but they often cost $50,000. "Our world was about computers that
cost $500," Gates said, and in that world, "we innovated right off the bat and always
made it take advantage of the machines."

The combination of speed and features in a program as small as Microsoft
BASIC was impressive indeed. Kemeny and Kurtz called the microcomputer
dialects of their language "street BASIC." But even they displayed some grudging admiration for the work that Gates and Allen did in 1975, calling it "a
remarkable achievement," before they added "but disastrous to the BASIC language." Today Kurtz looks back, recognizing the severe hardware restraints of
the early microcomputers, but he observed, "What was lost was part of the simplicity of Dartmouth's version of BASIC." In its messy transition from a teaching language to a working language, though, BASIC helped create the modern personal-computer industry.

In the spring of 1975, Gates and Allen signed a contract with Ed Roberts,
the president of MITS, located in Albuquerque, New Mexico. In the contract,

the pair referred to themselves as "Paul Allen and Bill Gates doing business as Micro-Soft." Long discussions about the company name had come back to the obvious – combining the first half of the words "microcomputer" and "software." The MITS contract was crucial, because its Altair machine was the early leader in the fledgling microcomputer business. But Gates, who dropped out of Harvard, had a more ambitious plan. He kept his independence from MITS, and made sure his contract did not limit his freedom to sell versions of Microsoft BASIC to other microcomputer makers. Gates wanted to sign up as many computer makers as possible on his way to making Microsoft BASIC a software standard. Selling wholesale to manufacturers, rather than retail to hobbyists, he was convinced, was where the money was – at least the money that could be collected. "Hobbyists liked our BASIC," Gates recalled 25 years later. But, he added with understatement, "They seemed to prefer to 'borrow' it from one another. They felt that it was unfair that we were asking for money."

Back in 1976, Gates was more vehement in his disdain for "software piracy," delivering a sarcastic attack in a famous "open letter" to hobbyists on February 3, 1976, published in the newsletter of the Homebrew Computer Club, a Silicon Valley mecca for hobbyists. "As the majority of hobbyists must be aware," he wrote, "most of you steal your software. Hardware must be paid for, but software is something to share. Who cares if the people who worked on it get paid?" He added that he "would appreciate letters from any one who wants to pay up."

His missive did not bring a flood of checks from hobbyists, nor did Gates expect that it would. He made his point, but he had already moved on to concentrate on selling his BASIC to personal-computer makers. Though young, Gates found he had an innate business sense, even as a salesman. He went to Texas to sell his BASIC to the Radio Shack chain for its TRS-80 personal computer. At Radio Shack, he met with John Roach, a vice president who later became chairman. Gates told him the price was $50,000, to which Roach replied, "Horseshit!" Gates was a bit taken aback by this Texas bargaining tactic. But, he recalled, "I held my ground, arguing that software was a crucial part of what Radio Shack's customers would need for their personal computers. Roach was a formidable guy, but he gave me my price." With each encounter, Gates gained confidence. "When it came time to sell to Texas Instruments, I decided that $100,000 was fair," he said. "But I was afraid that they might balk at six figures. So I offered them a grocery store bargain: only $99,000! It was a deal, and they bought it."

Microsoft's was by no means the only microcomputer version of BASIC on the market. In 1976, Gordon Eubanks Jr., a student at the Naval Postgraduate School in Monterey, California, wrote CBASIC. Eubanks had shown a skill for programming as a student at Oklahoma State University in Stillwater. Before he went to the naval school in California, Eubanks worked for IBM in Tulsa for six months. During that time, he wrote a program for Shell Oil that identified late payments on Shell's gas-station credit cards nationwide, and automatically sent out dunning notices to the people who held delinquent accounts. "It was solving real problems with this amazing technology," Eubanks recalled. "I really felt the power of that, and I found it absolutely addicting." Eubanks's CBASIC, selling for $100 a copy, did well among the small group of hobbyists-turned-professionals who were trying to make a living by writing specialized business software. But Gates was focusing on the more lucrative market of PC makers, working closely with them and convincing them to let Microsoft handle their software development. "We were very responsive to every hardware manufacturer's needs," Gates recalled. "We didn't want to give anybody a reason to look elsewhere. We wanted choosing Microsoft software to be a no-brainer."

Over the years, Eubanks would at times compete with Microsoft, and at other times cooperate with Microsoft. To Eubanks, it is no mystery why Microsoft fairly quickly emerged as the software leader. "Bill was, and is, a great business mind," he observed.

For Eubanks, CBASIC was a night project, working until 1 A.M. or so, then sleeping and getting up in time for his Navy studies each morning. He had a helper on this nocturnal venture, Alan Cooper, who, among other things, wrote the manual for CBASIC. "Without Alan's help," Eubanks recalled, "I never would have gotten it out the door." Cooper and his partner, Keith Parsons, had good reason to be interested in Eubanks's efforts. His CBASIC had a middle layer of code that made programs written in his language much more difficult to copy, modify, or steal. That was important to Cooper and Parsons, who were starting a company whose first product was a general ledger program, one of the early business programs for the microcomputer. His ambition, Cooper recalled, was to someday make $50,000 a year.

Cooper came from the hobbyist-inventor wing of personal computing. A high school dropout in 1969, he took to the road, traveling through

California, Alaska, and Western Europe aimlessly and, he said, doing "some serious self-medicating." A picture from that time shows a grinning Cooper, with
hair halfway down his back, standing next to a van adorned with psychedelic
designs. By 1972, he had tired of life as a dropout vagabond, got his high-school
equivalent degree, and enrolled in a community college in Marin County,
north of San Francisco. He intended to study architecture, but then he took a
class in computer programming.

"After a few weeks in that class, I was hooked," recalled Cooper in his office
in Palo Alto, where the one-time hippy now heads his own software consulting firm. "I began to devour computers and programming," taking every data-
processing course offered and working in the school's computer center.
"Programming is the perfectly controllable medium," he explained. "You can
do anything you want with code. You have absolute control. . . . I never did
study architecture." So, when the microcomputer revolution began, Cooper
plunged in, first in partnership with Keith Parsons and later on his own.
Cooper's mentality was that of the lone tinkerer at his software workbench,
inventing and solving problems. "I was scratching creative itches, moving from
one project to the next," he said. He found enough projects to make a living,
but not much more.

In the late 1980s, Cooper had come up with what he regarded as a neat tool
for "power" users – the dedicated hobbyists who loved to putter with their PCs,
but were not computer professionals. His program, called Ruby, was a "shell
construction set." It presented the user with a rectangular "slate" on the screen
– a kind of workbench for modifying the graphic shell of an operating system.
With Ruby, the user could, using a mouse, drag and drop items into the slate
and combine them – "like a set of digital Lego blocks," Cooper explained. It
enabled the technically adept user to fine-tune and modify programs, or combine two tools together. And Ruby's desktop "slate" could serve as the launching pad for a user's favorite applications.

Cooper soon decided that the logical place to take his invention was
Microsoft. In February 1988, Cooper went to the corporate campus outside
Seattle and showed his prototype software to Gabe Newell, one of Bill Gates's
young lieutenants. Five minutes into the hour-long demonstration, Newell
raised his arms and cut Cooper off, saying, "Bill has to see this." The meeting
was scheduled for the following month, a delay during which, Cooper said, "I
coded like a crazed weasel to put neat stuff in." Gates was impressed, and closed

the lengthy demo by telling his team, "We want to move forward with this." Microsoft purchased Ruby from Cooper, who chose to sell his software but not join Microsoft himself. "If I had, I could own a big chunk of Palo Alto by now," observed the hobbyist-inventor. "But I have no regrets."

The original idea was to make Ruby part of Windows 3.0, the version of Microsoft's operating system that shipped in 1990. Since the first version was marketed in 1985, Windows had struggled. The company, to be sure, had a formidable cash cow in its aging DOS, or Disk Operating System. But with its antiquated command-line interface – users had to type text instructions to start applications or display a list of files – DOS's days were numbered. In the end, Microsoft chose not to include Ruby, a tool for sophisticated users, in Windows. At the time, Microsoft's graphical operating system trailed well behind Apple's Macintosh in terms of being easy to use, and including a do-it-yourself, geeky tool like Ruby probably would not have helped the Windows cause. Cooper and some people within Microsoft, Gates recalled, had pushed for Ruby as a "shell customization tool" for Windows. But, he said, "that turned out to be a dead end. People did not want to customize the shell that way."

Yet Ruby was not abandoned by Microsoft. It would have a second life, being reincarnated as a programmer's tool. A big problem for Windows had been that businesses were slow to write applications that ran on Microsoft's graphical operating system. The problem was magnified by the fact that BASIC was being nudged aside as the leading programming language for personal computers by C at the time. But C, with its Bell Labs computer science lineage, was a language best suited for real experts, especially when it came to the complicated task of writing graphics applications in C. So, the number of people who could write Windows applications in C was limited. Somehow, Microsoft had to find a way for more people inside the corporate world to write programs that ran on Windows. Gates decided that the way to do that was to marry Ruby with BASIC. Ruby would be a visual tool, bringing point-and-click computing to programmers, that would sit atop Microsoft's Quick BASIC 4.0, a much-altered, turbocharged version of the venerable programming language.

Meshing the two was no easy job, but after nearly a year and half of development work, Visual Basic was introduced in May 1991. Visual Basic helped make Windows the industry-dominant operating system, and it helped the personal computer make huge inroads into corporate computing, tackling new jobs and taking over ones previously done on mainframes or minicomputers. Visual Basic made writing many business applications a matter of dragging icons into place on the screen, and writing a few snippets of code to stitch them together.

With a day or two of training, any COBOL or BASIC programmer inside corporate technical departments was suddenly a Visual Basic programmer. These corporate programmers mostly write software for specialized business tasks. They typically work in teams of one to five people, on projects that last from a few weeks to a few months. They are not developing operating systems that millions of people will use, but pragmatic, custom-made programs for in-house use, intended to save money or increase sales – programs that do everything from monitoring machines on the factory floor to measuring the effectiveness of advertising campaigns. Visual Basic opened up Windows programming to this vast silent majority of corporate programmers.

Alan Cooper, viewing Visual Basic from afar, was amazed at what had happened to the software he had sold to Microsoft in 1988. "The group at Microsoft did some great work, but I was pretty surprised," he said. "It's like sending your kid to college and he comes back summa cum laude, but he has had a sex change operation. It took some getting used to." Cooper made about $1 million in total from selling Ruby to Microsoft, and he gained industry recognition as the "father of Visual Basic." (Microsoft's lawyers once sent Cooper a cease-and-desist order, demanding that he stop using that title. But after Cooper complained, Gates patched things up and even lauded him as a "Windows pioneer" at an industry conference.) Looking back, though, Cooper is most impressed by Gates's strategic vision. "I thought I had written some pretty cool software," Cooper said. "But that it would become the programming control panel that is at the heart of Microsoft's success today, I could never have imagined. . . . That's why Bill Gates is what he is."

Bill Gates is not so much a creator of technology or a great programmer. His original BASIC was regarded as an exceptional piece of work – more an engineering feat than genuine innovation, though the distinction is a fine one in software. But Gates, most of all, is someone with a deep understanding of software, and the foremost applied economist of the past half-century. In the early 1980s, a group of young economists began studying the behavior of technology markets and the impact of industry standards, compatible products, and training costs on competition. They used terms like "network effects" and "switching costs," and at the time their real-world cases for examination included IBM in the mainframe era of computing, and the telephone system.

These economic forces, however, would work with particular force in software. It is a complex technical good, so "switching costs" for users are high. And software production costs – just replicating bits – are close to zero, once a program has been developed. Software markets were thus ones in which users would naturally gravitate toward widely-used industry standards, and a technology standard could lead to high profits and winner-take-all markets if the standard was owned by a single company. One of the seminal academic papers in this new branch of economics research was written by Michael Katz and Carl Shapiro, "Network Externalities, Competition and Compatibility," published in 1985. While they were working on their paper, Shapiro recalled Katz saying there was a guy who had been at Harvard when Katz was an undergraduate, who was doing precisely what they were writing about at a software company outside Seattle. He was speaking of Bill Gates, of course.

BASIC was the pliable tapestry with which Gates founded and first built Microsoft, applying the economics of software with a vengeance. It was not the destiny the Dartmouth professors had in mind when they created their teaching language, but its fate was determined in the anything-goes days of the early PC industry. As Alan Cooper recalled with characteristic plain-spoken candor, "BASIC was a whore of a language, and its main appeal was that anybody could bend it and twist it any way they wanted to for commercial purposes and for technical expediency." In the late 1990s, Microsoft's behavior in the software markets, when it was clearly the dominant firm, made it the defendant in a landmark Federal antitrust case. In June 2001, a Federal appeals court ruled that some of Microsoft's tactics to squelch any threat from Netscape Communications – the early leader in Web browser software – were an illegal use of its monopoly power, though the appeals court tossed out a lower court's order that Microsoft be split in two. Yet in the early days, when the company established its version of BASIC as an industry standard, it succeeded mainly because Bill Gates understood the economics of software better than anyone else, and Microsoft played the game better than anyone else.

In Visual Basic, some programmers have joked that the only thing left of the original BASIC is five letters. In a nod toward all the changes, Microsoft dropped the acronymic upper casing of the language in the name Visual Basic. "We have absolutely taken advantage of the fact that BASIC is a language we have total freedom to morph," observed Tom Button, the vice president of development tools. What Microsoft has done with Visual Basic since its introduction in 1991 is to play the technology standards game yet again, masterfully. By 2001, Visual Basic had become a programming standard used by a few million software developers, making it a leading programming language. For Microsoft, catering to developers – wooing them, helping them, supplying them useful tools, and hooking them on Microsoft's technology – is a corporate mission. The Microsoft mantra is, "A platform is an ecosystem." The developer community is a vital part of that ecosystem, and Microsoft feeds them assiduously.

Each year, a million programmers worldwide attend one of the hundreds of conferences or training sessions Microsoft sponsors for developers. There are a few thousand people at Microsoft whose job is to work with outside developers. "It goes back to the fundamental insight of Paul Allen and Bill Gates when they created Microsoft," Tom Button explained. "What did they do first? They created a programming language that enabled thousands of people at the time and later millions of people to write software applications. Attracting programmers is the crucial ingredient of success. That is the big lever in this industry. Microsoft was founded on that vision."

Visual Basic, like Microsoft itself, was built with the personal computer in mind. With the rise of the Internet and the Web, the personal computer will become less and less the center of gravity in computing. Microsoft understands this as well as anyone, and its response to the challenge is to try to establish a new programming platform on the Internet, which it has called .Net. A big part of that strategy is to woo all those Visual Basic programmers over to Microsoft's Internet platform. It is scarcely a surprise that Microsoft's Internet programming platform, Visual Studio.Net, bears a striking resemblance to Visual Basic. Microsoft's message to developers is that the language you know and the skills you have from the Visual Basic world will work fine for programming the Internet.

Standing in front of 6,000 programmers at Microsoft's professional developers conference in Orlando in the summer of 2000, Bill Gates emphasized that his company's approach to the Internet future would be much the same as the tried-and-true formula it has employed with such success in the past. We build a software platform, he told the audience, and we attract developers. "That has been the same since the beginning of time," Gates declared.

6

The European Influence: From Algol to Pascal to C++

THE MOST FUNDAMENTAL TOOLS OF SOFTWARE – programming languages – come in many varieties, covering a spectrum of styles and structure. But the mainstream languages, from FORTRAN and COBOL to Visual Basic and Java, do share a common geography: they were all created in the United States. America was by no means the only wellspring, but the programming languages that came from Europe, such as Algol, Simula, and Pascal, have been significant intellectually, though not commercially. A broad generalization, perhaps, but it does seem that the Americans typically brought an engineering mentality to the task of designing programming languages – compromises were made to solve the computing problem at hand. The Europeans, by contrast, often took a more theoretical academic approach to language design; the American mindset was more attuned to the marketplace.

The explanation is more nuanced than merely the obvious – that the United States was where the market for computing developed first, and the "animal spirits of capitalism" have always been found in greater measure in America. The practical bent of software development in the United States during the early postwar years was partly a product of other forces, and government largess played a big role. So much of the direction of computing in America at the time came from government and the emerging aerospace industry, working hand in glove, especially on military projects. At its peak in the 1950s, for example, the Pentagon's SAGE surveillance and air defense system alone employed 20 percent of IBM's workers – more than 7,000 people. The engineers and programmers working on projects supported by the Pentagon and aerospace companies tended to use computers to build real products – fighter jets, commercial airliners, and computer-controlled radar systems. "The Americans took an engineering approach because it made sense for them to use quick and dirty methods to achieve their goals," observed Martin Campbell-Kelly, a computer historian at the University of Warwick. The United States had both the money and most of the world's computers at the time. Naturally, then, the Americans were the ones were most likely to be grappling with the real-world challenges of trying to get the early machines to do useful things. The Europeans had neither that luxury nor that discipline, so a more leisurely theoretical approach took root there.

The United States–European distinction, whatever its precise causes, has eroded considerably in recent years. The United States is still the largest computing market, but it no longer has the dominance it once did. And the rise of the Internet has allowed ideas, and software, to jump instantly across borders and flourish in a global marketplace. Two of the most significant software developments of the past decade have originated in Europe: the World Wide Web was created in Switzerland by Britain's Tim Berners-Lee, and the popular Linux operating system by Finland's Linus Torvalds. Both were innovations of the engineering kind – tools developed by practical craftsmen who were initially just trying to solve some problem close at hand.

Even historically, the degree of international traffic in software ideas and techniques has not been fully appreciated. A brief look at the lineage of the C programming language is instructive. It was developed at Bell Labs, but it built upon BCPL created at MIT by Martin Richards, a British academic. BCPL stood for Basic CPL; CPL was a programming language jointly developed at Cambridge University and the University of London. Before the London team joined, the "C" stood for Cambridge; later, it officially stood for Combined, but unofficially it stood for Christopher, since Christopher Strachey led the development of CPL. Strachey later became the head of programming research at the Oxford University Computing Laboratory, and he was, appropriately enough, a firm believer in the benefits of mixing practical and theoretical work. "It has long been my personal view," he once wrote, "that the separation of practical and theoretical work is artificial and injurious. Much of the practical work done in computing, both in software and hardware design, is unsound and clumsy because the people who do it have no clear understanding of the fundamental design principles of their work. Most of the abstract mathematical and theoretical work is sterile because it has no point of contact with real computing."

Strachey, who died in 1975, would have been proud of Bjarne Stroustrup, a Dane who lives and works in the United States. Stroustrup took European software ideas, modified and refined them, and placed them firmly in the mainstream when he designed the C++ programming language, a descendant of Strachey's CPL. Stroustrup's work and career have been a merger of the practical and the theoretical. He was educated at Cambridge, where he got a doctorate, and he has spent his career mostly within the sheltered halls of AT&T's research laboratories. C++, Stroustrup says, was deeply influenced by his reading of philosophers like Aristotle and Kierkegaard and essayists like Albert Camus and George Orwell. Yet he is no dreamy theoretician, working in splendid isolation, when it comes to his craft.

Stroustrup was introduced to computers as a university student in Denmark, but he found the assignments dry and uninteresting – mostly programming mathematical proofs. Still, it was clear that programming came easier for Stroustrup than for other students. After his second year in school, he figured he had a marketable skill, and soon landed a part-time job writing programs for small businesses in Aarhus, a port on the Baltic Sea and Denmark's second-largest city.

The early working experience in Denmark would fundamentally shape Stroustrup's approach to computing. He worked for the local office of Burroughs, an American computer company that specialized in desk-sized business computers. As a contract programmer for Burroughs, Stroustrup went out and talked to the local businesses, listened to their needs, and then wrote original programs to solve their specific problems. It was made-to-order programming. He wrote accounting, billing, and payroll programs; his customers included lumberyards, mortgage companies, a gravestone maker and others. His programs were hand crafted in assembly code to husband the scarce resources of the Burroughs machines.

"I thought I knew something about programming," Stroustrup recalled. "I didn't know much, but it was enough to earn some money and learn some things. And one thing I could do was pack more into those machines than anyone else."

He became better at it as he went along, borrowing the best code from previous projects and refining it. The accounting system for the gravestone maker, for example, was a rejiggered version of a lumberyard program. He got particularly proficient at mortgage-calculation programs. By the time Stroustrup received his masters degree in 1975, more than one-fourth of all the mortgages in Denmark were calculated using his software. His part-time job in Denmark in the early 1970s, working with business customers face-to-face, left a lasting impression. He was struck by the difference between the academic work, often steeped in math and theory, and the real-world programming. "I always liked to build things and see things work," Stroustrup recalled in his suburban New Jersey office, overlooking a pond. "You can kind of get that with a math proof, but it's not the same thing. Software is something that you build and is used in the real world. And having an impact, doing something that improves people's lives, is important. . . . Being able to help that guy – it's something concrete, especially when that person's problem is shared by many others. And seeing that feedback is very motivating."

Those years in Demark fostered a pragmatic strain that would run through his later work. His upbringing also nurtured a practical mindset. Stroustrup grew up "solid working class," he recalled. His father Egon was a skilled workman – an upholsterer – and his mother Ingrid, a secretary, "the intellectual of the family," he said. As a computer scientist, Stroustrup is an intellectual, but one with plenty of grit on his hands. He is a practical philosopher who works in code. And his programming language, C++, combines the European skills in theory, design, and structure and the American flair for practical problem-solving, efficiency, and market savvy.

A lean runner and hiker, Stroustrup, 50, was wearing the standard uniform of informality found in the computing research culture – jeans, running shoes, and shirt with no tie – one day in early 2001 at his office in Florham Park, New Jersey, the home of AT&T Labs (the part of Bell Labs that went with AT&T when the company split up in 1995, while the rest of the lab went to Lucent). He has a close-cropped beard, a dry sense of humor, and reading tastes that run from weighty history and philosophy to Douglas Adams's *The Hitchhiker's Guide to the Universe* and Raymond Chandler's *The Long Goodbye*.

Stroustrup's C++ is a member of the extended family of Unix. It is a descendant of C, the language Dennis Ritchie wrote for Unix; C++ is a byproduct of a Unix research project. The assignment was to try to spread pieces of the Unix operating system across a network of smaller computers, all linked together. The project was intended as an early research foray into distributed computing, the long-sought goal of lashing many smaller computers together to build a powerful, resilient, and comparatively inexpensive computing environment. Stroustrup began working on the problem shortly after he arrived at Bell Labs in 1979, and soon decided that he would need simulation tools to analyze the network traffic that would result from splitting up and distributing Unix modules to many computers. So he began work on the needed software tools.

That proved to be an adventure of its own, resulting four years later in a new programming language, C++. "I never got to the project really," Stroustrup recalled, chuckling. "I got sucked into tools." This is, to be sure, a repeated pattern in software. A tool to solve a particular problem becomes a general solution. And the tools themselves take on a life of their own, and sometimes the original project is forgotten. It would be as if a person set out to build a house, before the hammer was invented, and then decides that something like a hammer would be a useful tool — and so designed a hammer. The person then decides that perhaps nails, and maybe a saw, would be neat to have as well. So mired in tool-making, the person never gets around to building that particular house, but the tools can be used by others to build countless houses. "What became C++ started just as a tool for expressing ideas better," Stroustrup observed. "And the set of ideas turned out to be interesting to a lot of people."

C++ would prove to be a programming power tool. It arrived in 1983, and it was a language well-suited for building bigger programs to handle the ever-increasing demand for complex software during the 1980s and 1990s. C++ gave programmers mechanisms to structure, define, and handle data logically. With it, a programmer can put one kind of data — say, employee information — in one kind of software container and then define what could be done with it, as if these controls were knobs or dials. A programmer can make a different software container for, say, scientific data, with different control knobs and dials on it for sophisticated math calculations. Then the data containers, or modules, can be assembled into related groups. Some containers can pass on information to kindred containers, while others cannot.

C++ helped programmers impose a real structure and logic on big programs when software, once again, seemed to be getting out of control. The more powerful computers of the 1980s could run bigger programs, but it became more and more difficult for software developers to build big programs – the same dilemma that precipitated the software "crisis" of the late 1960s. The bigger programs of the 1980s were costly to make, and their behavior was too often neither reliable nor predictable. Yet while C++ offers mechanisms for organizing and simplifying big programs, the programmer decides which ones to use. The programmer makes the rules in the software he or she builds. Unlike some other languages, C++ enforces few immutable laws. It provides instead a smorgasbord of options, an approach critics find needlessly confusing and treacherous for many rank and file programmers.

That obstacle, however, did not much hinder the spread of C++. It was embraced within the Unix community, and was often used for heavy-duty applications, such as running telephone switches. Like its C forebear, C++ enables a programmer to work close to the metal, allowing intricate control of a machine's performance. So C++ became popular in the industry for writing the software that handles tricky chores, such as controlling engine fuel injectors. Its flexibility made it useful in many different machines, often in hostile environments, which is why it was used by the Australian national lottery in its betting machines, many of them in the sun-baked outback. Stroustrup understood that most "computers" in the more distant future would not be machines with keyboards and screens. Yet, by the early 1990s, C++ had also moved into the center stage of present-day programming as personal computer software companies and others began distributing C++, including Borland, Microsoft, IBM, and Digital Equipment.

C++ is clearly the work of an individual, and in his *The Design and Evolution of C++,* Stroustrup explains his thinking with an intellectual sweep and with references one rarely – if ever – would find in another technical computer book. C++, he begins, was shaped as much by his "general world view" as it was by the computer science concepts that went into it. He has read history and philosophy for decades, which, he notes, has "given me a rather conscious view of where my intellectual sympathies lie and why."

"Among the long-standing schools of thought," Stroustrup writes, "I feel most at home with the empiricists rather than the idealists – the mysticists I just can't appreciate. That is, I tend to prefer Aristotle to Plato, Hume to Descartes. . . . I find Kierkegaard's almost fanatical concern for the individual and keen psychological insights much more appealing than the grandiose schemes and concern for humanity in the abstract of Hegel or Marx. Respect for groups that doesn't include respect for individuals of those groups isn't respect at all. Many C++ design decisions have their roots in my dislike for forcing people to do things in some particular way."

"Thus," he writes later, "C++ is deliberately designed to support a variety of styles rather than a would-be 'one true way.'"

The freedoms on offer are not for everyone. C++ is a language for serious programmers. Stroustrup's freewheeling democracy of software is an elitist preserve. A meritocracy perhaps, but there are few helping hands for beginners. "I want to give good programmers an advantage instead of protecting mediocre programmers from making all kinds of stupid mistakes," he explained in his New Jersey office. "I'm most interested in helping professionals build the things we depend on for our lives – telephone switches, money transfer systems and that sort of thing. So C++ does not have all kinds of protections against stupid novice mistakes."

Stroustrup's language grew up as a tool for grappling with the software challenges that came through the door at Bell Labs, and they were heavyweight problems that no novice would ever get near. Not surprisingly, the research problems frequently focused on the increasing complexity of telecommunications, so Stroustrup often worked on simulations of data traffic on big networks. Once he was asked to simulate the data traffic into New York's Manhattan Island, from the Wall Street financial district through the midtown corporate headquarters buildings to the uptown residential districts. At the time, he counseled that it couldn't be done. "But clearly, someone was trying to figure out the best way to wire Manhattan," he recalled. "There was a very practical side to the work."

Stroustrup has said that his "dream" programming language would blend of two European computing tongues, Algol and Simula. The story of Algol, in particular, sums up much about the European contribution to software historically – good ideas, undermined by poor packaging and implementation. Algol (short for Algorithmic Language) was a language by committee, whose major versions were published in 1960, Algol 60, and another eight years later, Algol 68 (Stroustrup's preferred version). Algol was the work of an international collection of computing experts, especially from continental Europe. Its intent was to lay the foundation for a formal syntax, grammar, and underlying logic in a computer language.

The Algol committee members viewed the dominant languages of the 1960s, FORTRAN and COBOL – both American creations – as a hodge-podge of useful features slapped together with a syntax that would look "nat-ural" to users – scientists and engineers for FORTRAN, and business managers and accountants for COBOL. And, of course, the Algol experts were right. By laying down basic language rules for computing, the Algol group hoped to cre-ate of formal body of knowledge that could be built upon and taught. They wanted to make computing a science and an academic discipline. As an exer-cise in pure language design, Algol 60 was a triumph. More than a decade later, Sir Antony Hoare, a British computer scientist, declared that, in terms of the principles of language design, "Here is a language so far ahead of its time that it was not only an improvement on its predecessors but also on nearly all of its successors."

But the Algol work was dominated by people who had very little experi-ence in hand-to-hand combat with the machine – the practical side of com-puting. Maurice Wilkes, former director of the Cambridge computer labora-tory, observed that in 1960 there were few powerful machines working in continental Europe. So the Algol group, unlike their American counterparts, were "not used to struggling with the day-to-day problems of getting work through a computer center. Their interest was more theoretical . . . They were determined not to be influenced by implementation difficulties for any partic-ular machine and, if a feature in the language made efficient implementation difficult, then so be it." Since it made few concessions to reality, Algol 60 ran so slowly that it made few inroads.

Another Algol effort was mounted later in the 1960s with similar results. Niklaus Wirth, a Swiss computer scientist, was a member of the later Algol group beginning in 1964. He recalled "endless discussions" of language philos-ophy, definition methods, and syntactic details. Two factions soon emerged. The "ambitious members," he said, wanted to explore new frontiers in language design and "erect another milestone similar to the one set by Algol 60." But the pragmatists, like Wirth, wanted to build on Algol 60, remove its deficien-cies, and add useful features, including a few from COBOL. The pragmatists lost, and in 1968 Algol 68 emerged with many strengths, but it was a complex language. Getting it compiled and running on machines proved a headache, so it was not in the marketplace until the early 1970s – theoretically pure, but ignored by most users.

Frustrated, Wirth set out to create his own language, a close cousin of Algol called Pascal, after Blaise Pascal, a seventeenth-century French philosopher and mathematician who invented a calculating machine regarded as a forerunner of the digital computer. Like Algol, Pascal was a language that adopted a structured design with blocks of code and clear definitions of different types of data. Yet Pascal, published in 1970, was designed more as a working language than Algol, stripping out the impractical theory and adding useful tools. Wirth's experience in America certainly had an influence. Though he created Pascal at the Swiss Federal Institute of Technology, Wirth got much of his computer education in the United States. His Ph.D. came from the University of California at Berkeley, and he taught computer science for four years at Stanford University before returning to Switzerland in 1968.

Wirth had plenty of hands-on experience wrestling with computers. He believed theory should serve utility. "I do not believe in using tools and formalisms in teaching that are inadequate for any practical task," he said. Indeed, a lot of serious programming was done in Pascal, especially for the Apple Macintosh. And in the early 1990s, Turbo Pascal, an extension of the original language developed by the Silicon Valley software maker Borland, became popular for a while among software developers as a programming tool for making applications that ran on Microsoft's Windows operating system. Still, Pascal's main influence was as a teaching language in universities into the late 1990s. It was the language for teaching structured programming, especially in serious computer science programs. In that way, Pascal has left a lasting impact.

Wirth's *Algorithms + Data Structures = Programs*, published in 1975, is considered one of the classics of computer science. Students reared on Pascal are a different breed than those who learned to talk to computers in BASIC, long the dominant language in the personal computer industry. "Pascal forced people to think clearly about things and in terms of data structures," said Philippe Kahn, a student of Wirth's in Zurich who went on to found Borland. "In BASIC, you think about individual commands with line numbers. So BASIC programmers jump all over the place, writing spaghetti code."

"Wirth's influence is extremely deep," he said, "because so many of the people who were taught in real computer science programs learned Pascal. It was the language of classical thinking in computing. There is a difference between a person who listens to classical music all day and a person who listens to rap all day."

"Of course, there is great rap music," added Kahn, who is an accomplished jazz musician. "But my point is that Pascal shaped the way generations of the best programmers think, and that is why the Algol-Pascal track is so important."

Algol can be thought of as the computing equivalent of Latin, a classic language that is at the root of other languages and is also used to teach language structure and grammar. Stroustrup would nod toward Algol, borrow some of its ideas, as he acknowledges, but he certainly wasn't going to use Latin. He was far too pragmatic. So he chose C, the language offspring of the Unix culture, a practical tool for programmers working close to the machine. Yet for his complex simulations at Bell Labs, he wanted other tools as well. For those, he borrowed ingredients from Simula, a programming language created in Norway in the 1960s. Simula was developed for studying the flow of things through networks – people through train ticket booths, products through manufacturing lines, ships through harbors, data through computer programs. It was the work of two scientists at the Norwegian Computer Center in Oslo, Kristen Nygaard and Ole-Johan Dahl, from 1962 and 1967.

Simula grew out of a discipline known as "operations research," which championed the application of statistical methods such as simulation to analyze and solve organizational problems. On paper, it seemed a neutral technique for improving everyday efficiency by smoothing traffic flows and the like. But in practice, there was a whiff of social engineering to this branch of management science. Human activity was an operation to be researched, and once understood, controlled. Perhaps not surprisingly, for a while the Soviet Union became an enthusiastic user of Simula, running on the government's Ural mainframes.

In Norway, Nygaard grew concerned that his creation was being used to organize work to the detriment of workers. The result, he observed, was "more routine work, less demand for knowledge and a skilled labor force, less flexibility in the workplace, more pressure." Nygaard's political sympathies lay with the Nowegian trade unions, leaving him with "a moral dilemma," he said. "I realized that the technology I had helped to develop had serious consequences for other people, especially the people that I had come to identify with politically." Nygaard went to the powerful Norwegian trade unions and helped them push for "data and technology agreements" with corporations and the government that gave the unions a say in the introduction and use of computer technology.

Stroustrup was interested in the technology of Simula, not its politics. The "classes" he saw in Simula were categories of data types, and indeed, the language that became C++ was initially called "C with Classes." Besides classes, Simula had the related concepts of inheritance and objects, which enable the programmer to compose software out of well-defined modules. A programming language, as a "natural" language like English or French, is a medium for expressing ideas, and Stroustrup found that Simula provided an elegant framework for expressing complex programming ideas clearly.

The Simula concepts help the programmer to label different kinds of data, organize the data in logical hierarchies, and then let knowledge flow from one data class to another related data class. For example, the programmer might create a class called "employees." Each member of the class is called an "object," so Mary Smith is an object, for programming purposes. The programmer will declare that each class member, or object, has certain attributes (name, age, Social Security number, salary, etc.). There are certain things, called "methods," that can be done with or to each class member (say, hiring, firing, increasing salary, or changing a medical benefits plan). There can be subclasses – "manager" might be a subclass of the "employee" class, for example. That new class of objects can inherit the characteristics of the parent class. So the programmer is only required to specify what makes "managers" different, while the general characteristics of employees are inherited.

This kind of programming with data objects arranged in logical hierarchies is known as object-oriented programming, and Simula was the first object-oriented language. And while Simula itself was always a niche language, its influence was far-reaching because it introduced the object-oriented technique to programming. Later, Alan Kay and others at Xerox's Palo Alto Research Center took the concept further with Smalltalk, an object-oriented language that helped bring point-and-click graphic computing to work stations. And Java, the popular Internet programming language, is object-oriented software.

Stroustrup's C++ was the vehicle that really delivered the Simula concepts to the programming community at large. He had initially encountered Simula in Denmark in the early 1970s, but he became really intrigued at Cambridge, where he was trying to do research on the software problems created by computing across networks. He was stymied by the lack of tools, and he figured Simula might be just what he needed. "You had to simulate back then to study distributed computing," he recalled. "You couldn't go out and buy a dozen computers and link them together. It was too expensive."

Since the late 1940s, the Cambridge University Computer Laboratory has been a European pillar of research excellence in computing. In 1949, Cambridge got the first stored-program computer operating, and a few years later its work on software "subroutines" established a model for the efficient reuse of blocks of code. The Cambridge Lab was founded by Maurice Wilkes, who was its director until 1980; he and Roger Needham, who succeeded Wilkes, conducted the admission interview of the young Dane. Stroustrup recalls the hour-long grilling as a grueling, yet exhilarating experience – being quizzed on programming languages, operating systems, and machine designs by a pair of leading computer scientists. To Stroustrup, it felt as if he was being worked over by an academic tag team. "It was tough because while one is asking questions, the other one is thinking up what to ask you next," he said. Stroustrup was impressive enough to be admitted in 1975 to the graduate program at Cambridge, where he received his doctorate in 1979.

Cambridge in the late 1970s, Stroustrup remembered, was a place where computing was in the air, pondered and discussed inside classrooms and socially as well – in the same way that one thinks of computing being part of the atmosphere at universities like Stanford and MIT. Stroustrup was able to get his hands on Simula as a result of a chance conversation at a Cambridge pub. One day, he was having a beer at the "Grad Pad," a riverside pub that is part of the graduate student center. At the bar, he recalled, he overheard some guy bemoaning his lack of success in getting Simula working. It was news to Stroustrup that Cambridge had an implementation of Simula anywhere. Simula implementations – the language, its source code, and a compiler designed to run the software on a particular machine – were expensive, costing $20,000 or so. Indeed, one reason Simula did not spread more widely was that the budget-conscious Norwegian Computer Center viewed its software research more as a product that should pay for itself than as knowledge to be distributed – so different from the stance that AT&T, for its own reasons, initially adopted with Unix. In 1973, for example, Donald Knuth was interested in experimenting with Simula at Stanford, but the Norwegians refused to give big price discounts to universities, and Stanford decided against acquiring a costly Simula license.

The frustrated Simula user at the pub was doing research elsewhere in the university. He was a stranger to Stroustrup, but he was generous with his software. "He said, 'I'll give it to you, if you can make it work,'" Stroustrup recalled. Soon Stroustrup had it up and running. Aided by Simula, Stroustrup was making swift progress on his Ph.D. With its logical structure of objects and classes, Stroustrup found the language a great help in building his complex computer-network simulation program. In Simula, he explained, a big program "acted more like a collection of very small programs than a single large program and was therefore easier to write, comprehend and debug."

But the smooth sailing did not last long. His Simula program was running on the university's IBM 360 mainframe, which served all the departments at Cambridge, including administration. While Simula had many fine features, efficient performance on the machine was not among them. It just chewed up the available data-processing cycles and memory of the IBM 360. "It was a pig," Stroustrup recalled smiling and shaking his head. "It could bring the mainframe to its knees."

Having the university's computing capability disabled by one student's research project was not something that pleased the administration. Stroustrup was told in no uncertain terms that the situation could not continue. He determined that the problem stemmed from fundamental features in Simula, such as its inefficient methods of handling memory and error checking. The trouble could not be fixed by tinkering and fine-tuning the software, so he abandoned Simula and was forced to start over, rewriting his simulation research in another language on a different machine.

Stroustrup painstakingly rewrote his simulator program in BCPL, the precursor to C that was designed by Martin Richards for Project MAC at MIT. After his stint at MIT, Richards returned to his native England to teach at Cambridge, where BCPL was often used on research projects. Stroustrup redid his program so that it could run on an experimental machine in the Cambridge computer lab that was not much in demand. Unlike the IBM mainframe, no company stood behind the experimental machine to help with routine maintenance, parts failures, and technical support. "But at times, deficient support can be a virtue," he said. "It keeps the amateurs off." He found the experience of coding and debugging his simulator program in the stripped-down BCPL to be "horrible."

Stroustrup had little choice, however, since his Ph.D. depended on his simulation program. His professors found him impressive, not only for his intellect, but for his energy. Roger Needham recalled Stroustrup working 100-hour weeks to get his program to work with a primitive language on an experimental machine. "He was a determined man who would let nothing stand in his way,"

Needham said. The experience left a lasting impression on Stroustrup, and the creation of C++ was motivated partly by the lesson he learned at Cambridge. "Upon leaving Cambridge," Stroustrup explained in *The Design and Evolution of C++*, "I swore never again to attack a problem with tools as unsuitable as those I had suffered while designing and implementing the simulator."

After Cambridge, Stroustrup would have preferred to return to his homeland, but he saw no interesting computer-science jobs in Denmark in 1979 – which would not be true today, he says. Still, it would be difficult to imagine a better offer for a computer researcher than the one he received from Bell Labs. "It was basically, 'You come and we'll give you excellent equipment and you work with nice people,' " he recalled. "'And in a year, you tell us what you did.' "

In hindsight, it seems, Stroustrup used that freedom so effectively because of his previous experience and his practical bent. Shortly after he began his tool-making project at Bell Labs, his intention was to combine the objects, classes, and inheritance of Simula with the efficiency of C. The idea was that C++ could be used for anything C was used for. It could be dropped into any C software project, and it would work without rewriting code. For years, the main compiler for C++ was a program Stroustrup wrote called Cfront, which translated C++ code into C code. That meant C++ would run on any computer that ran C, instantly opening a large potential market for Stroustrup's language. The C code produced by Cfront could be inscrutable, even to C programming experts, but it ran fast and reliably – testimony to Stroustrup's technical wizardry in service of the practical.

Stroustrup's goal was to improve things by building on the past. Even the C++ name signifies its evolutionary intent. In the C language, the "++" symbol means "increment," or add one. Depending on the context, it can also be read in C as "next" or "successor." Stroustrup borrowed even for the name, a suggestion from Rick Mascitti, another Bell Labs researcher. But it was a shrewd marketing decision nonetheless, suggesting a natural evolution instead of a disruptive revolution. "It was absolutely inspired," observed Douglas McIlroy, a former Bell Labs manager.

To fit in comfortably with the world of C, Stroustrup made plenty of trade-offs in C++. He consciously decided, he said, not to remove the "dangerous" or "ugly" features of C. So C++, for example, includes C's pointer mechanism, which gives a programmer unfettered access to memory – a sharp tool, but an often troublesome one. And C++ does not include a "garbage collection" utility to automatically find and sweep away chunks of data that are in the computer's memory but no longer being used. Some critics have called the lack of a garbage collector a grave oversight; it is a standard feature of Java, for instance.

Stroustrup notes that C++ allows garbage collection, and that add-on garbage collectors are widely available. First, he offers the philosopher's explanation for his design decision: "There is a difference between allowing and requiring garbage collection." Then, the pragmatic explanation: C++ was first used for a lot of low-level, next-to-the-metal work, like controlling fuel injectors. Programmers often shun garbage collectors for such work, because they can cause unacceptable pauses as the automated software tool runs through its routines. For these specialized tasks, professional programmers prefer hands-on coding to a sometimes-pokey automated helper. C++, Stroustrup says, would have been "stillborn," if he had included garbage collection as a standard feature.

Stroustrup made compromises in C++, but he did so with a clear-cut goal in mind: to bring the tools of structured language design and object-oriented programming into the mainstream. He took those ideas from Algol and Simula, fine-tuned them, and put them in a pragmatic language. Because of its C lineage, C++ seemed familiar to working programmers. And because of Stroustrup's skill, C++ was a fast, efficient language that could run on all kinds of machines. In all, C++ was a feat of making shrewd choices and some "very farsighted engineering," observed Brian Kernighan, a former colleague at Bell Labs who is a professor at Princeton University. "Certainly, it can be argued there are warts and bumps in C++. But if you go for what's ideal, no one uses it. If Bjarne had not done it the way he did, we wouldn't be having this conversation."

At AT&T Labs, Stroustrup holds the title of fellow, and is the head of the large-scale programming research group. He routinely declines job offers from companies offering considerably more money, especially from nearby Wall Street investment banking firms that have a voracious appetite for computing power and plenty of large-scale programming challenges. That more lucrative path, however, holds little appeal. "I like doing research that has an impact," Stroustrup said. "If I went to a company to make what they call 'real money,' I'd be just trying to make a system work as fast as possible to meet the product and service deadlines."

The practical research that appealed to Stroustrup in early 2001 was a project intended to find ways improve the reliability and resilience of software. If successful, it could help reduce data-traffic delays in telephone-network switches or help eliminate the problem of cell-phone service fading in and out when a customer is on the move. The incubator for his software research is a model motorboat. It has three tiny onboard computers, two engines, solar panels, a radio receiver, a local area-network, a tiny water cannon, and a weather station. It is heavy enough that it takes two people to carry it to the pond a hundred yards or so from Stroustrup's window. Stroustrup controls the vessel from his office from the notebook in his lap; each onboard gadget responding instantly to his commands, in theory.

On his blackboard, alongside the algorithmic descriptions of some of the boat's software, Stroustrup has written in bold letters, "It's a software project, dammit." It seems, he explains, some members of his research group were focusing too much on the model-boat side of the project – selecting the brass fittings, improving the finish of the wood, determining engine specifications. Boat, he says, is merely an acronym for Basic Object Architecture Testbed. "It just exists to exercise the code, to do cool things and see how it breaks," Stroustrup said.

7

A Computer of My Own:
The Beginning of the PC Industry
and the Story of Word

THE RUSSIAN-MADE URAL II WAS A HULKING BEAST OF A MACHINE. It consumed an entire room and was programmed largely by hand-setting switches on a mechanical console that resembled a turn-of-the-century cash register. A sea of tiny orange lights blinked behind the machine's glass doors and cabinets, each light denoting a circuit. Its electronic life pulsed for its human attendants to see, a pointillist symphony of bits. When, at 15, Charles Simonyi was allowed to set up and run a simple test program, "I almost fainted with delight," he recalled.

One Saturday, he got his chance to write his own program for the Ural II at the Central Statistical Office in Budapest. He tried to program a "magic square" – a math puzzle in which the rows and columns in a grid of numbers are all supposed to add up to the same number. Simonyi recalls setting the programming puzzle up as a 50 × 50 grid – an ambitious undertaking, given the circumstances. It proved a daunting task – a program with thousands of instructions and infested with bugs. He eventually got the program to run, filling in the correct numbers, and he came to view computing as a fascinating, compelling puzzle. He loved the feeling of control that came from mastering the machine.

To the teenage computer whiz in communist Hungary, the Ural II offered an intimate computing experience. In 1964, to be sure, it was a primitive machine by Western standards, resembling American computers of the early 1950s. But the Soviet computer did allow Simonyi direct access to the machine, not one of the arms-length styles of computing common in the United States at the time – time-sharing in which many people at terminals shared computing time, or batch-processing, in which a programmer handed his punched cards to a computer-center operator, who alone handled the machine. "As far as I was concerned, the Ural was a personal computer," he explained. "I was the only user on the machine when I was on it. I was very close to the machine. Every bit on that machine was mine."

Indeed, Simonyi's early experience – tactile, close to the machine, setting switches by hand – has strong echoes of two different computing generations: the pre-FORTRAN programming of the 1950s, and the first wave of microcomputer programming in the 1970s. In both periods, it was programming for the relatively few – computer-center specialists in the 1950s, and hard-core hobbyists in the 1970s. The arrival of the microcomputer, restyled invitingly as the "personal computer," was a fresh start for computing, a reinvention of sorts. It was made possible by a hardware breakthrough not initially intended for the computer industry – a programmable chip, or microprocessor, designed by the Intel Corporation's Ted Hoff for a Japanese producer of desk calculators. Yet electronics hobbyists – the kind of people who built their own radios from Heathkits – soon realized that the microprocessor could be used as the engine of a small, stored-program computer.

The result, at first, was that the room-sized computer of the 1950s was reinvented as the bread box-sized microcomputer of the mid-1970s. The early microcomputers were about as easy to use, and were programmed much the same way, as the big monsters of the 1950s. For the early hobbyists, the pain of getting the machine to work was part of the excitement. They were a small, inbred community, helping each other get their kit machines to run any programs, which were mostly simple games at the time. There was no real money in it, but as the chips became more powerful, it became apparent that these inexpensive machines were more than toys. They could do the work of *real* computers, and potentially put the power of computing into the hands of ordinary people. The personal-computer revolution that followed owed much to a seemingly unlikely combination of entrepreneurial energy and computer science. The personal computer industry was begun by outsiders – the hobbyists and young entrepreneurial upstarts like Steve Jobs of Apple Computer and Bill Gates of

Microsoft. But to bring computing to the masses, they had to borrow and build on the computer science that came out of universities and research laboratories. The early programming language of choice was BASIC, developed at Dartmouth College, and it was later joined by C and C++, both products of AT&T's Bell Laboratories. The way people interact with the modern personal computer – using a mouse to point and click, moving and manipulating visual icons on the desktop screen – can be traced back to work done at the Xerox Corporation's Palo Alto Research Center and Stanford Research Institute.

The life and career of Charles Simonyi traces a remarkable arc that spans computing periods and cultures, from the technological Dark Ages to the personal-computer era, from computer science research to the entrepreneurial payoff. He left Hungary at 17 and fled west, first to Denmark and then to California. He was educated at the University of California at Berkeley and Stanford University. He worked at Xerox PARC in 1970s, where he developed Bravo, a pioneering visual text-editing program. He joined Microsoft in 1981, and guided the technical development of some of the company's leading products, including Word (based on Bravo), which is one of the most widely-used programs in the world. Others, certainly, have made far greater contributions to research, but Simonyi was a member of the Xerox PARC team whose work shaped the course of personal computing for the next quarter century. Simonyi's efforts on Bravo alone were a significant contribution to the field – one that would eventually affect how millions of people create documents on computers, from business memos to novels. Simonyi is not an entrepreneur himself; he did not found a company, he is not a business executive, and his name is not widely known. But he certainly helped build Microsoft, especially in the early years. He has been an architect of some of the company's major products, a key member of the technical brain trust, and a recruiter of programming talent and researchers. Simonyi is a special programmer in whom the technology and economics of software converged in fairly spectacular fashion, who left communist Hungary penniless and became a capitalist billionaire.

Even as a child, Simonyi saw things differently, taking in the world around him with a cool analytic eye. He was an Erector-set fanatic, but parts were often scarce and rarely came in complete sets in East-bloc Hungary. Despite the shortage, he did manage to cobble together a "car" with a four-speed gearbox. "My car didn't look like a car, but it worked like a car," Simonyi recalled. "I was

doing abstractions in Erector sets." Once complete, he had no interest in playing with his mechanical creations. The appeal was in the problem solving; the
Erector-set parts were just the medium. "I was never going around z-z-z-z-z-z-z
with the car," Simonyi said, moving his hand quickly over the floor as if holding a toy car in an imaginary race. "I was weird."

When he was 14, Simonyi read a column in a Budapest newspaper that made
a lasting impression. It was written by an Italian communist, Antonio Gramsci,
and translated into Hungarian. As Simonyi recalls, the Gramsci article was a
parable about firewood and the humanity of communism's great leader of the
past, Lenin. As the story went, an old woman in the Russian countryside was
running out of firewood during a bitter winter. Supplies were short, and her
wood pile would not last until spring. Desperate, she wrote a letter to Lenin
himself, describing her plight. According to the parable, Lenin read the old
woman's letter and personally saw to it that she got a fresh supply of firewood,
thus saving her life. It was intended as a comforting tale of socialist benevo-
lence, but young Simonyi found it appalling. "I read this and thought, 'So
whether you freeze to death or not depends on whether the leader reads your
letter?' This was a system I wanted to flee as fast as I could."

His escape route, ironically, would pass through the Communist government's statistics agency, for it was there that he was introduced to computers
and programming. As a teenager, Simonyi was incessantly curious about
mechanical things, and he would attend trade shows in Budapest, inspecting
machines to "give me ideas," he recalled. Mostly, the shows featured things like
milking machines, tractors, and jackhammers. But he had heard about computers, read about them and wanted to get on one. Only the government had
them, but his father Karoly, a physicist and university professor, had a former
student who was the chief engineer at the Central Statistical Office. So Zoltan
Zsombok let the teenager in, took him under his wing and patiently explained
the basics of the Ural and of coding. The talented engineer in his twenties was
Simonyi's first mentor. Zsombok, who had asthma, would die of respiratory
failure in his thirties, but he had a permanent influence on Simonyi. "I think I
got this sense of how everything can be looked at as a computation from him,"
he observed.

Simonyi is dismissive of the tasks that his programming labors were put to
at the statistical office – doing calculations for raw materials and product ship-
ments under the guise of centrally-planned scientific socialism. "They were the

kinds of problems that having market prices would have solved in a millisecond," he said. Yet it was a wonderful laboratory for a bright, eager young programmer. He found the Ural to be an abstraction machine, often exhausting and frustrating to program, but marvelously versatile as well. "In a way, I found my ultimate Erector set," Simonyi noted, "an Erector set without limits."

His first professional software was a simple programming language for the Ural called Colur (Code Language for Ural). It was 5,000 lines long and written in the primitive vernacular of octal, the base-8 programming system, but it did enable a programmer to write arithmetic statements in code – understandable to the programmer, but not ready for calculation by the machine – that were then translated into a form that the Ural could execute, or run. Because of its translation program, or compiler, Colur was a programming language of a kind. Its value was as a time-saving tool for programmers at the Hungarian government agency. With its octal vocabulary, Colur may not have been pretty, but it worked. "It was my FORTRAN," Simonyi said with pride. In his safe at his lakeside home in suburban Seattle, he has no cash, no jewelry, no stocks or bonds. Instead, his safe holds just two items – his naturalization papers declaring his citizenship in the United States, and the paper tape of the first practical program he wrote, Colur.

That Simonyi chaffed at the collectivist restraints of communism is scarcely surprising to anyone who knew him later. Highly individualistic, Simonyi never fit neatly into categories. At Xerox PARC, for example, he was known as someone with an intriguing past, broad-gauged interests, and a taste for some of the showy toys of the West. He drove a Jaguar XKE sports car. "Charles was never a nerd with thick glasses," recalled John Shoch, a Silicon Valley venture capitalist who was at Xerox PARC with Simonyi. "He was a guy with a lot of different gears, even then." In appearance Simonyi, who has deep-set eyes, a boyish face and an angular nose, is often said to bear more than a passing resemblance to Napoleon, though taller and trimmer.

Simonyi entertains at his home, travels, and socializes, but by all accounts he has never been a joiner, and always a bit of a loner. He is a lifelong bachelor, but for years he has been a friend of magazine entrepreneur Martha Stewart. When her company first went public on the stock market in 1999, Simonyi was granted the largest allocation of "friends and family" shares. Simonyi's home is an extreme modernist statement; its interior is visually stunning, but most people would find living in such an environment intimidating, if not cold. All the surfaces are clean, polished; there are no plants, photographs, magazines, or anything just lying randomly around. "It's a real problem. No one can give me gifts and I have to be careful about what I buy," Simonyi observed with a smile, suggesting that he finds the aesthetic purity worth any problems it may pose. His home is truly *his* castle, with scant thought given to accommodating other people. A friend and colleague, Butler Lampson, noted, "Only Charles would build a twenty thousand-square-foot house with one bedroom."

Simonyi's exit from Hungary began with a small act of individualistic defiance: he grew his hair too long and was sent home from school. It occurred at the start of his junior year in high school. That evening, his father came home and asked him what he wanted for his birthday. Simonyi replied that what he most wanted was to not go to school any more. Instead of ordering him to fall into line, get his hair cut, and go back to school, Karoly Simonyi explained calmly that there were two ways to be educated outside the classroom in Hungary. The first possibility, he said, could be if a student had severe learning disabilities. The other alternative, his father explained, was to be determined gifted in some way.

By then, Simonyi's computer talents had been amply displayed at the government statistics agency. So, with his father's help, he petitioned the education ministry to be able to study at home, take school exams at twice the conventional rate, and thus finish high school a year early. As presented to the government, the special arrangement would free up Simonyi to pursue his computer studies. He worked at home, did the course work on the accelerated timetable, and studied English as much as he could. Since he could not begin his mandatory military service until he was 18, Simonyi had a year to work with, and he had a plan. At a trade fair in Budapest, he approached some engineers from a Danish computer company and quizzed them about their machine. He used that information to create what he called a "live resume" – a paper tape with a sample of his programming – and passed it to a representative of the Danish company at a later trade fair. "I gave him the tape, and basically said, 'Take this to your leader,' " Simonyi recalled.

A few months later Niels Ivar Bech, the director general of the Danish company, came to Budapest on a business trip and invited the teenager to lunch. Simonyi came prepared; he had studied the Danish Gier computer and its programs. That kind of information flowed more freely back then, and the machines and programs were less complex. "This was in the innocent times when it was possible to know everything about a machine and its software," Simonyi recalled. The Danish computer executive was suitably impressed, and shortly afterwards Simonyi received an invitation to work in Denmark for a year, and a plane ticket to Copenhagen. The Hungarian authorities allowed him to leave, but only after he had taken his university entrance examination and been admitted. In communist Hungary at the time, a university education was a prize obtained only by a select group, who were assumed to be set for life in professional careers. Apparently, the authorities figured that, with a university place waiting and his family all in Hungary, the teenager would return to Budapest on schedule, first serving his required military tour and then attending college in Hungary.

Simonyi, of course, had no such intention. His parents knew of his plans to stay in the West, but no one else did, not even his younger brother Tamas. A year later, after Simonyi failed to report on draft day, the military police came to the door, asking where Simonyi was. As Simonyi relates it, his mother, Zsuzsanna replied, "Well, I think he's somewhere between Copenhagen and San Francisco." His flight from Hungary had consequences; his father Karoly lost his post as a professor at the Technical University of Budapest. Later, however, Karoly Simonyi wrote a cultural history of physics that was a minor classic of its type and has been through several printings and translated into German, *Kulturgeschichte der Physik.* "It wasn't great that he lost his job, but he did his great life's work because of losing it," Simonyi said. "My Dad lived his life and he felt that I should live mine. He was very clear about it. He never gave me reason to regret what I did. He never would have done what I did, but he helped me do it."

Simonyi arrived in Denmark on Sunday, July 17, 1966 with no money and few possessions. His Danish boss gave him 500 kronor, equivalent to about $100 at the time, and took him out to lunch and dinner that day. Simonyi started work the next day at A/S Regnecentralen, a company jointly owned by industry and government that was intended to promote computer technology in Denmark. He was the third member of a programming team, and at first his new colleagues seemed skeptical about the Hungarian teenager. He began writing code on his first day on the job. "That broke the ice," Simonyi noted. "They decided I was their kind of guy."

In Denmark, Simonyi was introduced to the structured programming principles of Algol. While Algol was notoriously slow on most machines, it actually ran quite efficiently on the Danish Gier computer, thanks to a compiler program written at Regnecentralen. The Gier Algol compiler was designed by Peter Naur, a renowned Danish computer scientist who also helped Simonyi immigrate to the United States. Naur sent the personal letter of recommendation that surely helped Simonyi – a Hungarian with no high school transcript, who had taken no College Board exams – get accepted at the University of California at Berkeley.

At Berkeley, Simonyi majored in mathematical engineering instead of computer science. "I didn't have to learn computers," he said. "I was a professional." If he needed a credit or an "A" to prop up his grade-point average, though, he would took advantage of a Berkeley program under which a student could "challenge a course" – just take the final exam, without having attended any classes. Simonyi would walk into a computer science exam, and walk out smiling.

Simonyi naturally gravitated to the Berkeley computer center as a means of support. The pay at the university computer center was $2.95 an hour, but the work was by no means all routine, and Simonyi's labors there proved to be his tryout for the big leagues of computing. One day a young Berkeley faculty member, Butler Lampson, asked Simonyi to write a compiler for transforming programs written in Snobol, a symbolic list-processing language developed by Bell Labs in the early 1960s, into machine code. By then, Simonyi had considerable experience with the design and programming tricks needed to make a fast, nimble compiler, dating back to his days on the Ural, and it showed in the compiler he wrote. Lampson, who would become a leading software researcher, was impressed with the code and with the young Hungarian programmer. They would work together for most of the following decade, including collaborating on the creation of Bravo at Xerox PARC.

Lampson had been one of the lead engineers on Project Genie, time-sharing system at Berkeley that was backed by the Defense Department's Advanced Research Projects Agency. The goal of Project Genie was to build a computer system for small-scale time-sharing – a government-funded prototype to show that such systems were technically and economically feasible. It was a miniature version of MIT's mammoth time-sharing project, Multics. Using a large mainframe computer, the Multics project was intended to link

300 users to the machine at a time. The Project Genie team took a smaller Scientific Data Systems 930 machine and enhanced it with innovative hardware and software, yielding a time-sharing computer that could handle no more than 20 users at once. While Multics foundered, Project Genie was a technical success. Fresh from their triumph, the Project Genie alumni decided to take a big step ahead in time sharing. In 1968 they founded a private company, the Berkeley Computer Corporation, and made ambitious plans for a system that would cater to 500 simultaneous users. They raised money on Wall Street and began hiring.

Simonyi arrived at Berkeley in November 1967, too late to work on Project Genie. But he was there when Berkeley Computer was started, and his mentor, Butler Lampson, soon invited him to join. The Berkeley Computer machine turned out to be a classic example of Fred Brooks's "second-system" syndrome. Whereas Project Genie was shrewdly incremental in design, the BCC 500 system was a wish list. Berkeley Computer tried to do too much at once, and the company ran out of money and Wall Street's patience by the end of 1970. But Berkeley Computer gave Simonyi the opportunity to work closely not only with Lampson but also with Charles Thacker, a Genie alumnus and engineering wizard. Thacker, like Lampson, is a skilled computer systems designer , but one who tends to focus more on hardware. At Berkeley Computer, Simonyi wrote the microcode – a bottom layer of software that acts almost as circuitry itself, yet can be programmed, giving the designer the freedom to fine-tune the machine.

Even then, Simonyi displayed a certain flair. He would show up for the all-night sessions to eradicate bugs in the machine's software dressed in a "debugging suit" – a black net shirt and translucent, skin-tight black pants. But Simonyi was also the best debugger, Thacker said. "Without him, things would have gone much more slowly."

Throughout these years – indeed, until he received his Stanford Ph.D. in 1976 – Simonyi would hold what amounted to full-time jobs, as well as ostensibly being a full-time student. Simonyi has few doubts about which experience was more valuable, both personally and professionally. "Working with Butler and Chuck Thacker for a few years was worth more than a handful of degrees," he said. When he joined Lampson and Thacker at Xerox PARC in 1972, Simonyi was just finishing up his undergraduate degree at Berkeley.

Simonyi became a member of the elite group at Xerox PARC, but one without a bachelor's degree, a work permit or a green card, or permanent residence visa, which he would not obtain until 1974. "But hey, there was a desire for good programmers," he noted.

In the early 1970s, Xerox PARC was beginning its famed pursuit of a "dream machine" of computing – a small machine with a high-resolution screen used for work or play by a single person – a "personal" computer. To the children of the PC era, who grew up taking personal computers for granted, it is difficult to convey how revolutionary an idea this seemed at the time. Indeed, it was widely regarded as not merely impractical, but almost decadent – a waste of the costly and precious resource of computing. "The idea of one person having a computer – that was obscene," Simonyi said, recalling the prevailing view. But the visionary leaders at Xerox PARC, like Robert W. Taylor and Alan Kay, were fervent advocates of the concept of personal computing. Equally important, though, the engineers, like Butler Lampson and Charles Thacker, had come to believe that they could now translate the vision into reality. Rapid advances in semiconductor technology, storage, and displays, they concluded, were placing the goal of a personal computer within reach, at least in a research setting.

Thacker led the hardware design team that built the Alto, the ancestor of the modern personal computer. The first of the breed appeared in the spring of 1973. The machine had an innovative bit-mapped display that enabled each picture element, or pixel, on the screen to be manipulated by programming the bits in the Alto's memory. That opened the door to a previously unthinkable measure of visual flexibility by allowing a user to "paint" the screen with graphics and text of various sizes, shapes, and styles. The new hardware capability expanded the horizons for software. According to Simonyi, the bit-map technology brought a huge shift in the balance of power between hardware and software. "It meant that hardware does not determine what you see and do," he observed. "The hardware is just there to display bits. It is there to serve the software." In the early 1970s, he noted, the way most people interacted with computers was from an IBM terminal on a time-shared system. The IBM terminal screen was visual bread and water. It displayed numbers and letters, all capitalized, in the font IBM deemed appropriate. The Alto embodied a different approach, important technologically, and later economically. "It was a whole different mindset, a separation of concerns between hardware and software," Simonyi observed.

In early 1974, however, the Alto was still a computer in need of software to show what it could do. Butler Lampson figured that a program for creating and editing documents – a word processor, in today's terms – would be a good place to start. He was sketching out the rough design for a text-editing program one day when Simonyi wandered in his office. Lampson explained what he was up to and Simonyi, intrigued, glanced at the few sheets of yellow paper covered with algorithms and diagrams. Simonyi was interested in working on challenging projects at Xerox PARC, and especially ones that might serve the dual purpose of helping him get his Ph.D. at Stanford.

Simonyi has long been fascinated by the craft of programming and in ways to improve the process, tools, and techniques of programming. His Stanford doctoral dissertation was entitled "Meta-programming: A Software Production Method." Simonyi's meta-programming methodology had a number of time-saving features, including the use of "Hungarian," a consonant-filled notation shorthand to simplify the naming of data. Simonyi jokingly compares the appearance of his Hungarian to the Old English of Chaucer's time, and its name not only refers to Simonyi's homeland but is also a playful recognition that the strange-looking clusters of letters might as well be "Hungarian" for all the sense programmers can make of it when they first encounter the notation. Once learned, though, many programmers have found Hungarian an invaluable aid. Simonyi taught programmers at Microsoft to use his shorthand, and Word and Excel were written partly in Hungarian. The current edition of the *Oxford English Dictionary* includes Simonyi's notation method as one of the meanings of "Hungarian."

The essence of the meta-programming concept was to group programmers in teams; their work would be directed by the team leader, a meta-programmer. It embodied a kind of regimented division of labor in which the elite programmer would write detailed specifications for the subordinates who, working more as technicians, would write code. Meta-programming was yet another effort at "software engineering" – applying the engineering disciplines of strict job specification and process control to the craft of programming. At Microsoft, Simonyi flirted with the idea of installing a meta-programming regime, but soon abandoned the notion. It did not fit with the Microsoft culture of a small, fast-growing company trying to encourage each of its employees to behave like an entrepreneur. "We did use Hungarian and the tight team structure, which were elements of meta-programming," Simonyi said. "But we didn't try to organize it in a detailed way with cheap programmers and technicians taking orders from meta-programmers. That approach wouldn't have been right. . . . It wasn't like we tried meta-programming and rejected it. We didn't even try it."

Back at Xerox PARC, though, Simonyi was enthusiastic about meta-programming and eager to finish his doctoral dissertation on the subject. When he walked into Lampson's office that day, Simonyi had already planned one of his software-productivity experiments, using a couple Stanford students as underlings. The first experiment was to be called Alpha. To Simonyi, the Lampson text-editor project, residing as an embryonic concept on a few sheets of paper, looked like another test bed for meta-programming, and Lampson, with a little Tom Sawyer in his heart, encouraged the notion. "Charles's original motivation was not to write a text editor," Lampson recalled. "It was to try out his meta-programming ideas." So the project that would leave such a lasting impact on the software field began as meta-programming experiment No. 2, dubbed Bravo, the successor to Alpha. Before long, however, the text-editing software was the focus – far more than the meta-programming – but they kept the name.

Bravo was designed by Lampson and Simonyi. The collaboration was a matter of "constant discussion about high-level design," Lampson said, so that it was "very hard to say whose idea was whose." Simonyi was the programmer primarily responsible for making the ideas work in code. Bravo was a masterful sleight of hand, a program that made the most of the machine's resources by fooling it a little to perform its magic for the user on the screen. One of its clever ideas was an algorithm contributed by Lampson to stretch the Alto's memory resources when a document is represented by "piece tables." The piece table concept had been invented by another Xerox PARC programmer, Jay Moore. The algorithm allowed a document to be held in memory as a collection of text blocks ("a table of pieces") instead of representing each letter or character stored in memory bit by bit. If a writer wanted to take a paragraph from the middle of a short story and move it to the end, there would be three pieces: the document before the paragraph to be moved, the paragraph to be moved, and the remainder of the document. To the computer, in simple terms, the blocks $1 - 2 - 3$ become $1 - 3 - 2$. It was a matter of the shuffling of three blocks, or pieces, instead of many thousands of characters – the difference, in human terms, between having your groceries neatly packed in a few bags instead of trying to carry them loose in your arms.

Simonyi then made further refinements. He recognized that, in any given editing session, most of a document is untouched, and that separating the portion being edited from the untouched portion, each wrapped in its own software bundle, could yield further economies in computing resources. The piece tables, where editing changes were made, carry the "virtual document" seen on the screen by the user, but not stored in the document file until the editing session is over. "The piece table," Simonyi noted, "lets you edit without changing the file" – a further saving of computing resources.

Those "savings" would be spent in delivering rich text and graphics to the screens of users. Simonyi accomplished that by setting aside many bits per letter or character – 25 bits at first, and up to 100 bits later – to carry formatting characteristics of rich text images including bold, italics, underscoring, type size, font, and eventually color. "So you had space – space to grow for the next 30 years," he said. He then wrote the basic algorithms for encoding those properties with extreme efficiency, so that the Alto could handle it all. Changes in formatting would be made by "sending a message" – known as a formatting operator – to a piece of text of whatever size. All this was squeezed onto a machine whose processing power and storage capacity were minuscule by today's standards. The Alto's microprocessor ran at a speed of less than 6 megahertz, compared with the standard desktop computers in 2001, which run at 800 megahertz or higher. There was a lot of engineering artistry in Bravo. "To see what is needed is nothing by itself," Simonyi said. "You also have to see that it can be done and how."

Bravo combined insights, state-of-the-art algorithms, and close-to-the-metal coding – a testimony to Simonyi's multifaceted talent. His Bravo collaborator, Butler Lampson, observed, "Charles is just a phenomenal programmer at every level, from the bit level to the highest levels of software design." His other Xerox PARC colleague, Chuck Thacker, described Simonyi's achievement in Bravo as "taking a few good ideas from others, adding a few of his own, and building a system that had great conceptual unity," yet also ran efficiently on a machine of meager resources. "In a long career," Thacker added, "I've met very few people who can do this sort of synthesis nearly as well as Charles."

In designing Bravo, Lampson and Simonyi had concentrated on the underlying program, and it showed. At first, the user interface was clunky; it utilized what has become a pariah among man–machine interface designers, a confusion known as "modes." The user worked in either "text" or "command" mode, and any failure to keep the two straight could end in grief. The classic joke used to illustrate the peril of modes was the word "edit." In text mode, the four letters simply appeared on the screen. In command mode, however, typing "e" selected the entire document, "d" deleted it, "i" told the computer to insert what came next – so the hapless user saw all his or her work vanish from the screen to be replaced a single forlorn letter "t."

The answer to Bravo's problem was nearby. Two researchers in another part of Xerox PARC, Larry Tesler and Tim Mott, had been working for nearly two years on a text-editing user interface named Gypsy. They had made extensive studies of how human editors did their work at a text-book publisher in Boston. The methods and visual vocabulary of modern word-processing programs – "cut" and "paste" commands depicted with scissors and glue-pot icons, along with point-and-click mouse editing – come from their work.

Bravo met Gypsy in March 1975, to become Bravo 3. "The real winner," Simonyi noted. In the summer of 1975, a curious phenomenon began at Xerox PARC. Friends, relatives, and neighbors of the researchers began to come around to use the Altos to create and print announcements for Parent Teachers Association meetings, personal letters, college papers, business plans, and the like. For the most part, their presence was welcomed. They were human guinea pigs sampling the Office of the Future. "It may have been the first time that uninterested civilians were coming in to use computers," Simonyi recalled. "These weren't lab hackers or hard-core hobbyists. They just wanted to do something. That's when I said to myself, 'Wow, this is serious stuff.' "

The appeal was a fairly easy-to-use computer whose editing program could display on the screen text and graphics in all manner of type sizes and styles. The text and graphics could be printed out as a document just as it appeared on the computer screen – a capability taken for granted today, but breathtaking at the time. It became known as WYSIWYG, pronounced "wizzy wig," for a famous line uttered by the comedian Flip Wilson, cross-dressed as the wise-cracking, street savvy Geraldine Jones: "What You See Is What You Get."

The WYSIWYG editing that Bravo introduced to personal computing would help create the industry of desktop publishing. It changed the way professional editors, publishers, and graphic artists do their work, making it less laborious and improving their productivity. And desktop publishing has enabled millions of amateurs to produce their own newsletters, local magazines, and greeting cards. Bravo, like the Alto, was never a commercial product on its own. Even so, Bravo was the first "killer app" before the term had surfaced – a "killer" software application that made ordinary people want to use personal computers.

By 1976, the Alto had been recognized as a technology with potential at Xerox headquarters in Stamford, Connecticut. The Xerox plan was to "do the Alto right," which struck Simonyi as yet another case of "second-system" syndrome, or what his colleague Chuck Thacker succinctly termed "biggerism." In 1978, with its ambitious successor to the Alto, dubbed the Star, still years from the marketplace, Xerox set up a unit whose mission was to introduce the Alto to some big customers to give them a foretaste of the better world of office computing. The Alto, then, would be a poor man's placeholder for the Star. Simonyi gladly joined the new unit. About 1,500 Altos were produced, and they were used selectively in big companies like Boeing, and even in Jimmy Carter's White House. Simonyi's instincts about the Star proved accurate. When it was introduced in 1981, the Star was an impressive machine – powerful for its day, with a large, bit-mapped screen able to display text and graphics with remarkable clarity. But it cost more than $16,000, and seemed aimed at a lofty niche of the market, just when an inexpensive personal computer for business customers was being introduced by the most trusted name in computing, IBM.

During most of the 1970s, the Xerox PARC researchers generally regarded microcomputers such as the Altair 8800 and the Commodore PET, as a playthings for hobbyists and simple game machines for kids; they did not see them as *real* computers. These machines had none of the lineage of serious computing that came from the government, universities, and corporate labs. The people producing and programming microcomputers had no intellectual pedigree – or so it seemed to the Xerox PARC crowd. But by the end of the decade, it became apparent to some of the younger researchers at the Palo Alto center that the microcomputer phenomenon had the potential to change computing.

For Simonyi, the moment of realization came one evening after having dinner at a German restaurant in Silicon Valley with Paul Heckel, a consultant to Xerox PARC whom Simonyi had known for years. Heckel suggested that Simonyi might want to see a new software program that ran on his Apple II computer at home. Simonyi followed Heckel to his Palo Alto apartment and watched as Heckel demonstrated VisiCalc, the first electronic spreadsheet. "I was blown away," Simonyi recalled. "It woke me up. It showed that we weren't superior at PARC. VisiCalc was something really significant that we had not done at PARC."

After seeing VisiCalc, Simonyi became increasingly restive at Xerox PARC. He even briefly flirted with the idea of starting his own company. A hardware engineer at the Palo Alto lab, Bob Belleville, had made a "little Alto" using an Intel microprocessor with a bit-mapped display and a mouse. Simonyi visited Belleville to see the computer he had made in his garage. Enthusiastic, Simonyi suggested that they form a startup, with Belleville handling the hardware and Simonyi the software. Belleville considered it and balked, and Simonyi pursued it no further. In fact, Simonyi concedes, he had little real interest in starting his own company. He had always told friends his goal was to succeed in America without becoming a businessman. From the outset, he was confident his technical skills would make him wealthy. Chuck Thacker recalled one day at Xerox PARC. Simonyi drove up in his Jaguar XKE with a pile of camera equipment, costing thousands of dollars, all purchased on credit. "I pointed out to him that interest payments were an insidious pit in the U.S.," Thacker recalled. Untroubled, Simonyi replied, "No problem. I'm going to be rich, so this stuff will be easy to pay off."

Simonyi was not going to get rich at Xerox. Opportunity beckoned in the fledgling microcomputer industry, and Simonyi was going to chase it. He sought the advice of a successful Xerox PARC refugee, Robert Metcalfe, who created a popular standard for linking computers in networks called Ethernet, and then founded 3Com Corporation, a producer of networking equipment. Simonyi considered Metcalfe "the most savvy guy" he knew in the ways of the business world. Over lunch at an open-air restaurant at the Stanford Shopping Center in Palo Alto, Metcalfe gave him a piece of paper with the names of three or four people and the companies they headed. The three Simonyi remembered were Dan Fylstra, chairman of Personal Software, the publisher of Visi-Calc ("the top guy at the time," Simonyi said); Gary Kildall, head of Digital Research, which had an microcomputer operating system, CP/M; and Bill Gates of Microsoft. Steve Jobs of Apple, Simonyi says, may have been on Metcalfe's list, but by then Apple was a more established company, preparing to sell shares to the public in late 1980. Simonyi was looking for the added excitement, and payoff, of becoming an insider in a fledgling company. "Apple didn't have that charm," he recalled.

Simonyi started with Microsoft, and went no further. He met Gates while on a business trip to Boeing, which was trying out a few of the Xerox Altos. On his last day in Seattle, Simonyi drove to the Microsoft offices, whose entire operation occupied half of the eighth floor of a bank building in the suburb of Bellevue. Simonyi had thought Metcalfe had called Gates beforehand, saying that Simonyi would be dropping by about 2 P.M., but there was a mixup, and Metcalfe had not phoned in advance. "It was a completely cold call," Simonyi said, "but I didn't know that." Gates was meeting with a group from a Japanese company, so Simonyi was ushered in to see Steve Ballmer, Microsoft's second-in-command. Simonyi had brought some materials with him, including a consulting firm report that praised Bravo in glowing terms. While Simonyi's visit had been unannounced, his reputation had preceded him. "We had heard about Simonyi," Gates recalled.

Gates emerged from his meeting with the Japanese, but by then Simonyi was in danger of missing his flight back to San Francisco. So Gates drove Simonyi to the airport and the two talked "all the way to the gate" – about how personal computers would someday be in offices and homes everywhere, about Gates's plans for entering the software applications business with products like Bravo, and about how Microsoft was going to grow and grow. "I talk to this guy, and he solves all my problems," Simonyi said. "It's fantastic."

After Simonyi returned to California, Gates went to see him, and they talked for hours. Afterwards, Simonyi wrote a brief memo – three pages and a few paragraphs on a fourth – summarizing their conversation and amplifying his own thoughts. He wrote it in November 1980 on an Alto, using Bravo and an 8 ½ × 11-inch sheet of paper, folded over, like a church-service program. He gave it to Gates to make sure they were thinking the same way about Microsoft's plans and strategy before Simonyi joined. "Charles wrote a great document that was his synthesis of things we had talked about for a good 10 hours," Gates observed. "It was really a combination of Charles's vision and my vision."

Given the state of the industry and of Microsoft at the time, the memo – essentially a strategic outline with brief comments and asides – is an extraordinary document, considering what Microsoft has done since it was written. Back then, Microsoft was a supplier of software tools with fewer than 40 employees, and its leading product was its version of the BASIC programming language,

Microsoft BASIC. It had no operating system of its own, and no applications programs such as spreadsheets or word-processors.

The memo declared that Microsoft's business would be to produce and sell software for "the mass market" – not that one existed at the time. Under a category labeled "Strategies," it stated that Microsoft would have a "full product line" and a unifying "brand name," with "compatibility implied" among the products carrying the Microsoft banner.

A key strategic objective, the document stated, was "control of operating system/programming environment." Here, in particular, the historical context is what makes such an assertion so revealing about the long-term vision of Gates and his early recruits like Simonyi. In November 1980, Microsoft had privately signed a deal with IBM to supply an operating system – or master control program – for the big company's entry into the personal-computer business in 1981. Yet when it signed the crucial contract with IBM, Microsoft did not own an operating system. It had, however, agreed to pay $25,000 to a local company, Seattle Computer Products, for the right to resell its operating system, 86-DOS.

In July 1981, less than a month before the IBM PC was introduced, Microsoft paid Seattle Computer Products an additional $50,000 to buy 86-DOS outright. The transaction gave Microsoft its foothold and foundation in the operating system business. The often-told story of how the company got into the operating system business has become part of the Microsoft legend, seen in hindsight as evidence of Gates's underhanded skullduggery or of his gutsy entrepreneurial genius, depending on one's perspective. In any case, when Simonyi wrote his memo in November 1980, the strategic objective to "control the operating system/programming environment" reflected a measure of visionary premeditation that no one outside Microsoft appreciated at the time.

The 1980 Simonyi memo stated that Microsoft should organize its business in terms of two markets for personal-computer software: an office market, and a home market. Then, he wrote, "Is the idea of 'model year' applicable to software?" As he pointed to that sentence in the memo, which he keeps in his files at home, Simonyi said, "So does that sound like Windows 95, Windows 98 and Windows 2000?" Under a section on office applications, the memo mentioned that Microsoft should develop a competing product to VisiCalc. "That's Excel," Simonyi said, referring to Microsoft's spreadsheet program. The memo also mentioned the need for a database offering. "That's Access," he said, referring to the Microsoft product. And the memo mentioned the need for a product like Bravo. "That's Word," Simonyi noted. His memo then noted "mouse point-ing" and the point-and-click user interface that Xerox had developed and Apple was adopting with its "Apple IV" machine – later named the Lisa, and a predecessor to the Macintosh. Such a computing style, Simonyi wrote, "could be retrofitted to practically all micros, opening up a software replacement mar-ket." Indeed it would – as Windows, the Microsoft operating system, that now runs on more than 90 percent of all personal computers.

Bill Gates calls Simonyi "one of the great programmers of all time." But in Simonyi, Gates saw more than technical brilliance. He saw someone who shared his vision of software as something separate and special, economically, culturally, and technologically. "Our whole idea was that software was different than hardware, that the way we would run the company, the way we would hire people would be different," Gates recalled. "And Charles had that same set of beliefs, so we knew when we hired Charles that we were hiring one of the very key people to build the company around."

In February 1981, Simonyi joined Microsoft as the director of advanced product development, which meant software applications. "There were no applications," he said, "and I was to do it – the technical part." When Simonyi arrived, work was already underway on a spreadsheet program similar to Visi-Calc. Called Multiplan, it was character-based product with a few modest improvements over VisiCalc – a far cry from Simonyi's plans for putting snappy graphics, database capabilities, and other fancy features into a spreadsheet pro-gram. The forward-looking plans would be shelved for the moment. VisiCalc had been originally designed for the Apple II, and it had been slow to appear on other computers.

So the Microsoft strategy was to get its spreadsheet, Multiplan, running on as many machines as quickly as possible, exploiting the VisiCalc weakness. Simonyi had a clever programming strategy for achieving the goal. The Microsoft devel-opers would write so-called byte code – an in-between layer of software, which is not ready to be compiled to run on a specific machine, but for a software inter-preter. The idea was that Microsoft could write an application like Multiplan once, and then let the manufacturers write their own interpreter programs, con-verting the byte code to machine code for each different kind of computer. More than 100 manufacturers signed up for Multiplan, including IBM, NEC, Olivetti, Datapoint, Texas Instruments, Commodore, Digital Equipment, and many others. It seemed like a good idea at first – a broad strategy for a period of ferment, particularly in the microcomputer business. Yet Microsoft paid a price for its cover-the-waterfront approach. Multiplan ran slowly, especially compared with a new competitor, Lotus Development Corporation, which bet on the IBM PC. The Lotus spreadsheet, 1 – 2 – 3, ran on the Microsoft operating system in the IBM PC, but also sidestepped it at times to get improved performance and added features. "Lotus essentially hand-coded 1 – 2 – 3 for one machine, the IBM PC," Simonyi said. "It ran five times faster than we did. We got our asses kicked by 1 – 2 – 3."

The good news for Microsoft, though, was that the IBM PC had emerged as the industry standard, establishing Microsoft's DOS as the industry-standard operating system. Because IBM allowed its technology suppliers to sell to others, producers of IBM-clone machines were founded and flourished, including Compaq, Dell, and Gateway. But if IBM itself would no longer rule the industry, the suppliers of the IBM-standard technology – Microsoft in operating system software, and Intel in microprocessors – would become more and more powerful. Supplying the IBM-standard operating system did not guarantee Microsoft's success, but it certainly helped. Microsoft had to make the tricky transition to an operating system based on point-and-click mousing and graphic icons – a graphical user interface, or GUI (pronounced "gooey"). Indeed, the shift from DOS to Windows was slow, and at times unsteady for the company, and for the entire PC industry, partly because the hardware had to evolve too. Yet once Gates had recruited Simonyi – the emissary from Xerox PARC, the fount of graphical point-and-click computing – the Microsoft course was set for the long term. As soon as the GUI-based Xerox Star was introduced, Microsoft bought one of the costly machines "so our people could get a feel for it," Simonyi recalled.

Microsoft's applications division was also prodded toward the graphical computing by one of its biggest customers, Apple Computer, the commercial pioneer of GUI with its Macintosh computer introduced in 1984. The Macintosh would never really threaten to topple the IBM PC-standard computers as the market leader, but it showed people a better, easier way of personal computing. In the DOS world, the user began with a dark screen except for one illuminated item, C:\>, known as the "C prompt" or "DOS prompt." To call up a file, say, a business memo or letter, the user had to remember in what computer directory the file was stored, and then type DIR, followed by a space and then type the directory name. Next, to actually summon the business memo or let-ter onto the screen, the user had to type CALL, followed by a space and then type the name of the file. A forgotten directory name or a single errant key-stroke brought frustration. In the Macintosh world, by contrast, files were sum-moned visually, by point and clicking on icons or following a menu that dropped down from the top of the screen with a mouse click.

To ordinary computer users, the DOS environment seemed like program-ming. Inevitably, the vernacular of personal computing would shift from typed commands to pointing and clicking. Microsoft understood that and the transi-tion would give the company's applications division an opportunity to displace rivals – Lotus 1 – 2 – 3, WordStar, and WordPerfect – who at one time appeared to be firmly entrenched leaders. "The Mac saved our behind in applications because it was a sea change," Simonyi explained.

The principal method of technology transfer in the software business is people moving. The passage of Bravo into Word is a classic example. Butler Lampson had occasion to peer closely into the insides of Word after he joined Microsoft in 1995, and the Bravo parentage was transparently evident. "It uses the same data structures, the same architecture, the same ideas," Lampson observed.

The medium of software is malleable and thus often eludes tight, crisp legal definitions. The restrictions on people as idea messengers, carrying concepts and detailed technical knowledge from one institution to another, are not many. In the years since Simonyi went to Microsoft, the legal environment sur-rounding software has tightened considerably, yet even today, there would appear to be few curbs on the kind of borrowing and building on his previous work that Simonyi did in Word. "There is a conceptual link between Bravo and Word, and I did both," Simonyi said. At Xerox PARC, there were no contrac-tual prohibitions on pursuing such similar work elsewhere. At Microsoft, the standard employment agreement states that a programmer cannot work in the same product area with another company for a year after leaving Microsoft. But Simonyi adhered even to that standard, since he worked first on the Multiplan spreadsheet before moving on to design Word. In the current environment, lawyers would have been looking over his shoulder to ensure that no docu-ment or note he might have brought with him from Xerox could raise a legal question. Asked about any intellectual property issues surrounding Word,

Simonyi said, "We didn't touch anything written." Later, however, when going through some files, he pulled out a copy of the Bravo X technical plan, published in May 1980 by Xerox. "I probably shouldn't have it," he said shrugging.

The first version of Word was released in November 1983, with more graphics capability than either IBM-standard PC's or printers of the day could handle. Its underlying technology looked back to its GUI origins at Xerox PARC, and ahead to its GUI future, first on the Macintosh and later on Windows. Microsoft Word's sales would build gradually to become, after Windows, the world's best-selling software program – the program most people use when they write letters, business reports, student papers, newspaper articles, magazine stories, and books. The lonely artist, locked in a tiny room with visions of writing an Oscar-winning screen play or the Great American Novel, is these days most likely staring at a computer screen with his or her aspiring masterpiece rendered in Word – Simonyi's creation.

Simonyi was leafing gleefully through stacks of recently acquired recreational reading. They were copies of the operations and countdown manuals for America's Mercury, Gemini, and Apollo space missions, some of them signed by the astronauts. "This isn't a curiosity for me," he explained. "I want to read them." Simonyi's home is an intellectual playground, with its vast library, its modern art, its trove of engineering memorabilia, and its computer lab – a sleek white room with floor-to-ceiling white boards and brightly colored bean bag seats ("it's the PARC tradition," he noted, referring to the Xerox Palo Alto lab). His philanthropic activities are largely focused on promoting serious intellectual pursuits – the mathematics building at Princeton's Institute of Advanced Study bears his name, as does an endowed chair in science at Oxford University. When he visits Oxford, Simonyi stays with Richard Dawkins, a famed evolutionary biologist and holder of the Simonyi professorship. They rarely speak of business or politics, Dawkins says, but instead their discussions tend to focus on science, history, and culture. Simonyi, said Dawkins, is "a real intellectual, fascinated by all aspects of science, including its history, and deeply knowledgeable in many fields."

Among his other interests, Simonyi is fascinated by languages. Besides English and Hungarian, he has studied French, Danish, and German, at times having revolving tutors coming to his home. An appreciation of what languages do and how they work, he observed, is useful in thinking about programming languages, and how different they are. "There is this pernicious analogy that computer languages are really like human languages," he said. "But in fact, computer languages are far more restrictive of what you can say in various ways." Programming languages, Simonyi says, are less general-purpose than the term suggests, and far less so than human languages. FORTRAN was a language intended for science and engineering; COBOL for business; C for systems programming; Lisp for artificial intelligence, and so on. "Look at the difference," he said. "It's not as if German is for business and English is for science."

Since the mid-1990s, Simonyi has been working on overcoming what he regards as the stultifying limitations of programming languages. His ambitious goal is to transform the craft of code writing – deliver a Big Bang-style breakthrough in programming productivity. Many people are skeptical, and they regard Simonyi's pursuit as quixotic. Progress in software engineering, they say, has always been evolutionary. Over the years, new tools and techniques have brought steady progress, but software will remain something of a hand-crafted art, unlike the science of hardware. Gradual improvement is the best programmers can hope for, the skeptics insist, and theirs is certainly the prevailing view in computer science. In *The Mythical Man-Month,* which remains the defining treatise on the limitations of the craft, Fred Brooks declared that there is "no silver bullet" to the software productivity problem. "Complexity is the business we are in, and complexity is what limits us."

Simonyi begs to differ. "Everything Fred Brooks said was true, but it's irrelevant in a way because the context is so much different than it was 20 or 30 or 40 years ago" he said. Computers, he noted, are a thousand times more powerful than in decades past, opening the door on new software methods. "To say there is no silver bullet is really a cop-out."

Simonyi is working on a technology that he hopes will free the human intelligence of the developer – the programmer's "intentions" – from the confining syntax and conventions, the walled cities of individual computer languages. He calls his system "intentional programming," and it is a programming supertool for capturing the mental shorthand of abstraction – what people want the computer to do – in modules of code that can be used with any programming language. "I want to liberate the abstractions from these closed programming languages," he said, "and let programmers put their cleverness in an independent delivery vehicle."

The intentional programming system is complex, covered by patents held by Simonyi and others on his team. Yet what it does, essentially, is allow a programmer to work at a level above a conventional programming language, focusing on *what* the person wants to accomplish rather than *how* to program it. In computing terms, the *what* statements are declarative programming, while the *how* instructions are procedural programming, and traditional computer languages require a programmer to often labor in a dense undergrowth of procedural detail. Simonyi's new technology uses elaborate logic trees, linked to a vast database, to help assign the proper context and meaning to data rendered in software – that is, to help divine the intention behind the data. Literal-minded computers are notoriously blind to context, and Simonyi says that is largely because programming languages are so inflexible.

In Simonyi's scheme, the software ideas and features written in one language can be mixed and matched with the best features of others, the conversions made through the magic of the intentional programming development environment. Intentional programming, he says, is a "transformational" technology in the sense that it enables the programmer to work above the restrictions of conventional languages. "This is a move one transformational level away from the current paradigm, a step closer to human intentions," he said.

Only time will tell whether Simonyi's idea will deliver a breakthrough in software productivity, or prove to be a reach too far. If successful, it could wed even more software developers to Microsoft technology, adding to the company's wealth and influence in the industry. The man himself is characteristically optimistic. "How to save and reuse human intelligence of the programmer is the most basic issue in computer software. By far, it is the best problem.... I think it's going to be huge. I've been wrong before, but not about the big things."

Computing for the Masses: The Long Road to "Gooey" and the Macintosh

ANDY HERTZFELD LEARNED AN EMBARRASSING LESSON about the inflexible, literal-minded nature of computers in high school. He wrote a program for matching partners for the junior prom, but it matched one girl with a third of the boys in the class. To the teenage programmer, it seemed obvious that one girl could not go out with 30, 50 or 100 boys at a time. Yet the dating faux pas was not obvious to the computer, unschooled in human ways – which is to say unprogrammed to eliminate multiple matches.

The "prom bug" setback did nothing to cool Hertzfeld's enthusiasm for computing. A year earlier, he had enrolled in a programming course at Harriton High School in suburban Philadelphia, and he was smitten. He did his programming, mostly in BASIC, by remote control, from a Teletype terminal linked remotely to a machine he never saw, a big time-sharing mainframe system. In 1970, time-sharing was the only way high school students could gain access to a computer – if their school was sufficiently affluent to purchase precious time on some faraway, multimillion-dollar machine. The school paid for computer time by the minute, and soon Hertzfeld was single-handedly racking up bills at an alarming rate. "I was banned from using the machine," he recalled. He wormed his way back on with pledges of contrition and restraint, but mostly by proving to the school administrators that he could write useful programs, like one that calculated the grade point average and class rank for each student.

Almost from the beginning, whether solving math problems or devising dating schemes, programming appealed to Hertzfeld both intellectually and emotionally. "You're making stuff, stuff that wasn't there before and you have absolute control. When you are a kid, no one listens to you. But the computer does," he said, recalling the sensation of power over the machine. "It was something that resonated with my soul. I loved it. And I was good at it."

Hertzfeld programmed purely for enjoyment. As a teenager, he gave no thought to making a living someday by writing software. He thought of programmers as corporate drones – "people who had horrible jobs, wearing a coat and tie and working for a bank or someplace like that." Not an image to lure a talented high school student immersed in the youth culture of the time, which was largely defined by what it rejected – the Vietnam War and corporate America. Hertzfeld loved songs like Bob Dylan's *Highway 61 Revisited*, odes to the open road and the "rebel spirit." He attended Brown University, studying physics, math, and computer science, but the experience did not little to alter his dreary image of professional programmers. "Around Brown at time there weren't any good role models, at least not that I saw," Hertzfeld recalled.

Hertzfeld went to graduate school at the University of California at Berkeley, but without much enthusiasm. He found the computer-science courses to be dry, tedious, and soul-deadening. Then, in 1977, Hertzfeld encountered the Apple II and found his calling. Here was a machine that sat on a person's desk, and for its time was an engineering masterstroke – compact and affordable at $1,500, it still packed the punch to display color graphics and ran a solid microcomputer version of the BASIC programming language.

The Apple II was the first machine to give a real glimmer of what was to come in personal computing. It suggested a future well beyond the nerdy, hobbyist machines of the time – essentially Heathkits wrapped around microprocessors, most of which had all the visual appeal of a high school industrial-arts project. The Apple II, by contrast, was housed in an inviting plastic case, an early nod to computer aesthetics that bore the unmistakable imprint of Steven Jobs, Apple's cofounder. But the impressive guts of the machine – the chips cleverly chosen, arrayed and programmed for peak performance – showed the handiwork of the company's other cofounder, Stephen Wozniak. The more

Hertzfeld studied the Apple II and tinkered with it, the more impressed he was. Its engineering had a personality, an individualistic, even playful touch – a certain rebel spirit, as Hertzfeld saw it. "It was a real computer, but it wasn't about crunching numbers faster," he recalled.

Hertzfeld looked into the Apple II and saw the future he wanted to pursue: writing programs that made personal computers more accessible, more useful, and more fun for ordinary people. He quit school and began writing programs for the Apple II, mostly neat ideas rendered in software like the one that enabled the machine to display lower-case as well as upper-case letters. Hertzfeld was hired by Apple in 1979, and he would later program the working heart of the Macintosh, the machine that brought point-and-click computing, using graphic icons nestled on a virtual desktop, into the mass market. The Macintosh team would include a neurologist, a biochemist, and a graphics artist, all of whom, like Hertzfeld, caught the personal computing bug and believed they were changing the world for the better.

The Macintosh introduced a different look-and-feel into the mainstream of computing – a more friendly face, a better "user interface," in the geekish term of the craft. The people who determine the face that software presents on the screen must make judgments about programming, graphic design, psychology, and taste. They are engaged in the highest level of translation and communication between man and machine – that is, furthest away from the machine and in deep sympathy with the human user. The computer scientist who designs a programming language is also a translator, but for a narrower audience, typically other professionals. The user interface designer works on a broader stage, and must cater to a much larger, less sophisticated audience. The language designer is speaking to the converted; the user interface designer is trying to reach new converts.

User interface software requires different skills and a different mentality from much of computer science. It is as much humanities as engineering, and requires a different breed of person – someone like Andy Hertzfeld, a code hacker who cares deeply about look-and-feel issues. "It's really fun to be in the middle of the technical, precise and objective computer," he explained, "and the fuzzy, emotional, subjective human being. I've always loved art, especially literature and music, and I think the human element is what can elevate engineering to the realm of art."

Some 15 years after the Macintosh was introduced, Hertzfeld was back at it.

In 1999, Hertzfeld and a few members of the original Macintosh team founded
Eazel Inc., a startup that designed a user interface for machines running Linux,
a Unix-based operating system built using the open-source model through
which software is collaboratively developed and distributed free. Eazel folded
in May 2001, unable to get continued funding as the financial markets soured
and investors shunned untested new ventures. Short, elfin, and bespectacled,
Hertzfeld, whose business card since his Apple days has carried the single title
"Software Wizard," was disappointed, but his belief in the importance of making computing more accessible was unshaken. "To me, it's not a job," he said.
"It's a calling, to make things great for the people sitting in front of the
machine."

The quest to design computers and software that would be more friendly
to humans really began with J. C. R. Licklider. A Harvard-trained psychologist,
Licklider set up a psychology program at MIT in 1950. He decided to place it
in the electrical engineering department, hoping to encourage engineers to
design with human beings in mind. During the 1950s, he worked on the Pentagon's SAGE project and other air-defense surveillance systems that combined
radar and early computers. In such systems, rapid and clear presentation of
information was essential, and Licklider helped design the display consoles.
There were, he said, "two aspects of the presentation of information problem:
building the stuff and then getting it to work, making a good interface with
the user." Licklider later moved onto Bolt, Beranek, and Newman, a technology research firm in Cambridge, Massachusetts, where he wrote his seminal
paper, "Man-Computer Symbiosis," in 1960.

In his essay, Licklider stated that the appropriate goal of computing was to
increase the productivity of knowledge workers, to "augment" human intelligence rather than substitute for it. This may seem obvious, but at the time there
was considerable optimism among artificial-intelligence researchers that computers would match or surpass human intelligence in many realms of thinking
and problem solving before long. Maybe someday, Licklider said, but in the
meantime the big gains would come from making computers more useful assistants, freeing up people for the real thinking and letting the machines handle
the drudge work. And there was plenty of the latter, based on Licklider's
research. He estimated that 85 percent of a technical worker's time was spent
"getting in a position to think" by doing menial "clerical or mechanical"

chores. His man–machine symbiosis meant finding a way to shunt much of the humdrum 85 percent onto the computer.

There were, however, significant hurdles to the closer relationship. Communicating with the computer was at the top of the list. "The basic dissimilarity between human languages and computer languages," Licklider wrote, "may be the most serious obstacle to true symbiosis." He noted the progress of programming languages like FORTRAN and Algol, but mainly as evidence of humans bowing to the machine. "Men are proving their flexibility by adopting standard formulas of representation and expression that are readily translatable into machine language," he wrote, and then addressed the limits of that approach. "For the purposes of real-time cooperation between men and computers, it will be necessary, however, to make use of an additional and rather different principle of communication and control." He went on to discuss the need for better displays and input devices, and even pointed toward the distant technological shores of handwriting and speech recognition.

On its own, the Licklider paper was an important contribution to research, but he was soon in a position to have a more direct impact on the field. In 1962, Licklider was named director of the Advanced Research Projects Agency's new information-processing technology office. Licklider's unit of the Pentagon research agency was given a multimillion-dollar budget to subsidize research in computing, and the grant-maker naturally took a particular interest in human–computer interaction. Licklider's interest and research grew out of his own hands-on experience with computers, both in his work on air-defense systems and then at Bolt Beranek, which bought the first computer the Digital Equipment Corporation produced in 1959, the PDP-1.

Licklider's formulation that only 15 percent of a knowledge worker's time was spent on real thinking was based on logging and examining his own work habits one summer. His instinct, based on his own exposure to computing, was that the machines could truly change things. "I thought, This is going to revolutionize how people think, how things are done," Licklider recalled in 1988, two years before he died. "I was one of the very few people, at the time, who had been sitting at a computer console four or five hours a day – maybe even more. It was very compelling. I was very frustrated at the limitations of the equipment we had, but I also saw how fast it was getting better." Licklider was an early believer that information technology would yield a big economic payoff in productivity gains. "I thought we were going to double it or triple it, or multiply it by four or ten or something. I still feel that way," he said in 1988, speaking of the unrealized potential of computing.

Shortly after he arrived in 1962 at ARPA, the Pentagon research agency, Licklider heard from a kindred thinker in the field of human–computer interaction, Douglas C. Engelbart. A researcher at the Stanford Research Institute (SRI), Engelbart had just published his professional manifesto, "Augmenting Human Intellect: A Conceptual Framework." In terms that echoed Licklider, he noted the plight of the knowledge worker, faced with increasing complexity and a growing glut of information, and declared that the computer offered "the greatest immediate promise" of helping humans out.

In his paper, Engelbart sketched out his vision of how things would be for a worker in the computer-enhanced future. Using a keyboard and an advanced display console, a person would be able to manipulate text and symbols. After an initial draft of a document was typed onto the screen, the worker would "begin to edit, reword, compile and delete. It's fun – 'put that sentence back up here between these two' – and blink, it's done." The laborious physical grappling with pencils, typewriters, scissors, and glue pots would be abandoned, replaced by a computer-aided "new process of composing text." On the computer, "trial drafts could be rapidly composed from re-arranged excerpts of old drafts, together with new words or passages which you stop to type in. . . . You can integrate your new ideas more easily, and thus harness your creativity more continuously." The knowledge worker of the future would be amazed by the difference, and Engelbart even wrote of the elation this hypothetical worker might feel. "You reflected that this flexible cut-and-try process really did appear to match the way you seemed to develop your thoughts. Golly, you could be writing math expressions, ad copy, or a poem, with the same type of benefit." Anyone toiling as a "creative problem solver," Engelbart insisted, would be far more efficient working with a computer.

Four decades later, it is difficult to appreciate Engelbart's prescience fully. His futuristic predictions have become reality, even commonplace, but he was writing at a time when computers cost hundreds of thousands or millions of dollars. The notion of individuals having their own personal work stations seemed inconceivable. When Engelbart wrote, the cost of a screen display alone ranged from $20,000 to $60,000. The future that Engelbart envisioned in 1962 was not entirely on target. He assumed, for example, that the knowledge worker of the future would use a light pen as well as a keyboard. The light pen never did take off as a popular input device for manipulating characters and symbols on the screen. What prevailed instead was the computer mouse, which Engelbart invented a few years later.

Engelbart and his team at SRI, funded by the government, came up with a flurry of innovation during the 1960s that shaped how people interact with computers. Besides the mouse, the biggest step was being able to present multiple views of data on the screen, as side-by-side "windows." The concept of windows changed people's relationship to information. Instead of presenting information by the page, the computer window is a lens providing one view of a landscape of data, which would become known as cyberspace. Other windows can be summoned onto the screen to provide other views, other perspectives. Engelbart's inventions would be refined, modified, and added to, first at Xerox PARC, then at Apple, and eventually at Microsoft. But the modern face of computing began with Engelbart. Alan Kay, a leader at Xerox PARC, called Engelbart "a prophet of Biblical dimensions."

The seeds of Engelbart's career were planted in the Philippines. A 20-year-old Navy electronics technician, Engelbart was on the island of Leyte, waiting to be sent home at the end of World War II, when he wandered into a makeshift Red Cross library, up on stilts in a hut. Browsing through some magazines, he found a July 1945 copy of *The Atlantic Monthly* and read an article entitled "As We May Think," by Vannevar Bush, director of the Office of Scientific Research and Development that coordinated research for the war effort. With the war ending, Bush pointed to a new challenge for technology, one created by the "growing mountain of research." To solve the problem, he proposed a "memex," a mechanized desk with a screen and keyboard, which would store all of a person's books, records, magazines, and communications on microfilm. Storage, in Bush's vision, would be virtually unlimited. "If the user inserted 5,000 pages of material a day, it would take him hundreds of years to fill the repository." Engelbart found it stirring reading. "It just thrilled the hell out of me that people were thinking about something like that," he recalled.

Yet Engelbart would not translate the thrill into action until years later. He was working as an engineer on wind-tunnel experiments at the government's Ames Research Laboratory in Mountain View, California, but he was restless.

Pondering his future, Engelbart decided that "I wanted to invest the rest of my heretofore aimless career toward making the most difference in improving the lot of the human race." Recalling Vannevar Bush and reflecting further, Engelbart chose to commit himself to "augmenting the human intellect." In 1951, he enrolled in graduate school at Berkeley, earned his Ph.D in electrical engineering, and then joined SRI in 1957. His early efforts to lure government research funding for his work were disappointing. East Coast chauvinism played a role. In 1961, Engelbart recalled that one rejection letter stated "since your interesting research would require exceptionally advanced programming support, and since your Palo Alto area is so far from the centers of computer expertise, we don't think that you could staff your project adequately."

When Licklider arrived at ARPA the following year, the money began to flow. By 1964, the SRI team was experimenting with devices for marking and moving things on the screen. The mouse emerged victorious in a contest that involved rigorous testing in a tiny market sample within the confines of SRI. The mouse triumphed over the light pen and the joy stick, as well as more unusual alternatives like a device attached to a user's knee and a "nose-pointing control." The mouse, Engelbart explained, "consistently beat out the other devices for fast, accurate screen selection in our working context. For some months we left the other devices attached to the workstation so that a user could use the device of his choice, but when it became clear that everyone chose to use the mouse, we abandoned the other devices."

Recognition of Engelbart's achievements came slowly. In 1968, SRI staged an impressive demonstration of their work-station technology at a major computer conference in San Francisco. With Engelbart on stage, the images on the work-station were projected onto a 20-foot screen. Engelbart demonstrated the use of the mouse to manipulate interactive programs; he split the screen in windows and mixed text and graphics. The 90-minute performance was a virtuoso technology demo, involving a forest of antennas and microwave lines to link the convention center with the SRI labs 30 miles away. "Afterward we thought for sure that the world would be talking about everybody starting to augment now," Engelbart recalled. "Well, it didn't happen."

Over the years, Engelbart would often say that he was amazed at how long it took for the SRI work to make its way in the broader world of computing. Inertia and failure of imagination undoubtedly played some role, but so did economics and human nature. Computing at the time was still extremely costly, so the concept of the personal work-station seemed almost a theoretical matter – fascinating, but not practical any time soon. Don Chamberlin, who later created the SQL database language, was a graduate student at Stanford University when he attended an Engelbart lecture in 1969. Using the mouse, Engelbart moved images of ships around on his workstation screen. "It was neat and sexy, but I had no inkling that twenty or twenty five years later everyone in Western civilization would spend much of the day pushing a mouse around," Chamberlin recalled.

In some ways, Engelbart was proposing an uncompromising revolution, rather than more moderate evolution. His "augmentation" work-station was a comprehensive system, with certain elements more accessible than others. Years after his government funding ran out and he had left SRI, Engelbart still championed something he called a keyset device, which he originated at SRI. He used the keyset as a complementary input device to the mouse, one in each hand. The five keys could be struck in chords, like a piano, and required the user to learn a stenographic-style code. "Whenever you hear someone say it has to be 'easy to learn and natural to use,' put up a little flag and go question it. What's natural," Engelbart declared. "What's natural is what we've grown to accept."

To Engelbart, the desire to make things easy to learn and use was often the path to second-best solutions. He pointed to the tricycle, so much easier to learn than a bicycle; yet once trained, the bicycle rider had so much more speed and range. Engelbart did not shy from requiring humans to bend to the machine. The SRI mouse had three buttons.

At Xerox PARC, the terms of human–computer communication would shift markedly in favor of the person in front of the screen. During the 1970s, the Xerox Palo Alto Research Center created the modern metaphor for interacting with computers – a "desktop" arrayed with little pictures known as "icons," a mouse, overlapping windows, and simplifying "menus" for guiding the user through the confusing complexity of the computer. The Xerox PARC-style of computing centered around the "graphical user interface," or GUI (pronounced "gooey"). It also acquired another acronym, WIMP, which stood for its main elements – windows, icons, mouse, and pull-down menus – and hearty purists who felt cute little pictures had no place in *real* computing found the term particularly appropriate. The reactionaries, of course, were eventually routed and gooey took over, starting in the 1980s.

The Xerox Corporation opened its Palo Alto research center in 1970. Xerox could certainly afford the research outpost, thanks to the cash flowing from its copier business. But the Xerox management saw that its lucrative franchise would come under attack, as inexpensive Japanese copiers were already making inroads in the market for small machines. With that in mind, Xerox wanted to diversify beyond copiers in the long term, and the broadly-defined mission of the Palo Alto group was to search for the "Office of the Future." Its pursuit of that future was shaped by the thinking on man–machine interaction pioneered by Licklider, and the Xerox PARC work was done largely by researchers who were alumni of various projects funded by the Pentagon's ARPA unit.

The intellectual beacon and guiding spirit behind the user interface work at Xerox PARC during the 1970s was Alan Kay. A brilliant iconoclast, Kay was thrown out of one college for protesting its quota on Jews, made a living for a while as a professional guitar player, and then, during a stint in the United States Air Force, he was found to be a gifted computer programmer. After the military, Kay went to the University of Colorado, where he studied mathematics and molecular biology. He then went onto graduate school at the University of Utah, where he became immersed in computing in earnest. In the late 1960s, the University of Utah was the nation's leading center of computer-graphics research – another ARPA-supported program under Professors David Evans and Ivan Sutherland.

When he arrived at the University of Utah in 1966, Kay was handed a copy of Sutherland's classic "Sketchpad: A Man–Machine Graphical Communication System." Sketchpad was the program Sutherland wrote in the early 1960s for a TX-2 computer, one of the first machines with a visual display. Sutherland was graduate student at MIT, Sketchpad was his Ph.D. project, and it represented a striking advance for its day. It was the first drawing program at a time when computers were thought of as giant calculators. But Sutherland considered himself a "visual thinker," who once explained "if I can picture possible solutions, I have a much better chance of finding the right one." Why not enlist the computer's help in visual thinking? With Sketchpad, a person using a light pen could draw pictures, designs, and blueprints. The program enabled the user to modify, copy, and store the drawings, zoom in and zoom out on the display – experiment and test out ideas in a way that could not be done in the traditional medium of paper, pencil, and eraser. "What it could do was quite remarkable," Kay recalled, "and completely foreign to any use of a computer I had ever encountered." It was the invention of modern interactive computer graphics.

Almost immediately after encountering Sketchpad, Kay was introduced to another important influence, Simula, the simulation programming language developed by two Norwegians, Kristen Nygaard and Ole-Johan Dahl. Simula was structured in "classes" of data types, and embodied other concepts like inheritance. Simula allowed the programmer to label different kinds of data, arrange the data in logical hierarchies, and then let information flow from one data class to another related one – a system for simplifying complexity, but one with more flexibility than most programming languages. Simula got Kay thinking in biological terms, such as cell metabolism, with its "notions of simple mechanisms controlling complex processes and one kind of building block able to differentiate into all needed building blocks."

Just as Sketchpad was a new use of the computer, Simula was a different approach to programming languages. "It is not too much of an exaggeration to say that most of my ideas from then on took their roots from Simula," Kay explained later. "It was the promise of an entirely new way to structure computations that took my fancy."

The following year, Doug Engelbart visited the University of Utah to show off his system. To Kay, the Engelbart vision was "a compelling metaphor of what interactive computing should be like, and I immediately adopted many of his ideas." At the time, Kay also began pondering Moore's Law – the observation made by Intel's cofounder Gordon Moore in 1965 that the number of transistors crammed onto a computer chip seemed to double about every 18 months. If the trend continued, Moore reasoned, computer power would increase exponentially over time. To Kay, the implication of Moore's law suddenly seemed clear – the massive, costly computers of the day would be transformed into affordable machines that sat on desktops, or laptops. It was just a matter of time, and Moore's Law.

For Kay, it was a revelation. "Computing as we knew it couldn't survive. . . . It must have been the same kind of disorientation people had after reading Copernicus and first looked up from a different Earth to a different Heaven." The Copernican shift that Kay foresaw in 1967, just as IBM was rising to absolute dominance in the industry, was that the mainframe was destined to lose its place at the center of the computing universe. So, "instead of at most a few thousand *institutional* mainframes in the world," he foresaw that there would someday be "millions of *personal* machines and users, mostly outside of direct institutional control."

Over the next year, Kay would be influenced by other ideas. At a retreat for graduate students working on ARPA projects, he listened to a lecture by Marvin

Minsky of MIT's artificial intelligence lab, in which Minsky attacked the stultifying, lock-step nature of traditional education. For the first time, Kay heard of the theories of childhood development and self-discovery of Swiss philosopher and psychologist Jean Piaget, and of the work of MIT's Seymour Papert and his learning-by-doing research with a programming language designed for children. Kay visited Papert and observed his experiments with children, who built things and solved problems in a language and programming environment called Logo that involved manipulating on-screen agents that looked like turtles. Kay read Marshall McLuhan's *Understanding Media* and pondered its delphic epigram, "The medium is the message." He visited the University of Illinois, where he saw a one-inch square "lump of glass and neon gas in which individual spots would light up on command – it was the first flat-panel display," whose larger descendants can be found on millions of laptop computers today.

These ideas soon crystallized for Kay, clarifying his concept of the personal computer. Whereas Engelbart thought of his systems as personal "vehicles" – cars as opposed to IBM's mainframe "railroads" – Kay believed that the destiny of personal computing was to be a personal, interactive *medium*. The insight carried a series of implications, including how people learned to use computers. "With a vehicle one could wait until high school and give 'drivers ed,' but if it was a medium it had to extend into the world of childhood." But Kay did more than think. He modeled his idea into a mockup cardboard model of what the ideal personal computer should look like. Recalling that Aldus Manutius, decades after the invention of the printing press, determined the modern dimensions of the book by making sure it fit into saddlebags, Kay carefully considered the human dimensions of his dream machine. It would be carried with no more difficulty than a book, and rest lightly in a person's lap. It would have a flat-panel screen, a keyboard and a stylus (since it would recognize handwriting), and he poured in lead pellets to see how light it should be (less than two pounds). It would communicate using a wireless network. Kay called his concept computer the Dynabook.

The Dynabook, like Charles Babbage's Analytical Engine, was one of the most important computers never built. Yet the Dynabook would nonetheless exert a powerful influence in computing in precisely the way Kay described. In the wider world, the name "Dynabook" is unknown, but to computer people it has long stood for the idealized goal of personal computing. Companies are still chasing Kay's dream, almost exactly as he imagined in 1968, and they are getting closer all the time.

A machine, even a dream machine, is not a medium. Kay's vision of computing for the masses meant that software would have to transform the experience of computing. "Millions of potential users meant that the user interface would have to become a learning environment," he explained. To Kay, a new software environment meant starting from first principles – a new programming language. The trouble with computer languages seemed to be that they worked too much on the machine's terms – a rigid collection of procedures, data structures and functions – that had none of the flexibility of natural language. So Kay set out to design a new language, which he called Smalltalk. As it took shape during the 1970s at Xerox PARC, Smalltalk was not merely a language, but a fresh approach to organizing computation, a more flexible way to map human problems to the machine.

Kay replaced the machine-like architecture of traditional languages with a biological metaphor, in which the basic building blocks were "universal cells," or objects. Anything could be represented as an object – numbers, words, lists, and pictures – and they interacted by messages sent to each other. Each object in Smalltalk could be thought of almost as its own self-contained computer, a virtual machine. "It struck me that programming would be incredibly easier if we only had virtual machines to deal with," Kay recalled, just "as biological organisms are only constructed by cells." He also thought of Smalltalk as a software environment that would be much like the early Internet, the Pentagon-funded ARPANET just taking shape at the time, which was "lots and lots of machines communicating only by sending messages to each other."

Kay described this cellular style of building software as "object-oriented programming." Other languages use object technology – Simula before, and C++ and Java afterwards – yet none had the purity of conception, or have adhered to the object philosophy in the same way. Smalltalk, Kay said, is "still the only *real* object-oriented language," where everything is a bona fide object in a system that is entirely self-contained. The major "object-oriented" languages now in use, C++ and Java, made pragmatic compromises to run faster and to accommodate the habits and practices of working programmers. To Kay, a bold conceptual artist in the medium of computing, such compromises seem shortsighted. "They usually only go after a part of the problem, while congratulating themselves on being 'practical.'" Still, he takes satisfaction in seeing that developments like Java and Microsoft's Internet software initiative, called .Net, are emulating Smalltalk in some ways. The Internet is forcing the industry to think in terms of simplifying complexity by emulating Smalltalk's cell-like design. "Software in general," he observed, "is slowly getting more and more like Xerox Smalltalk of 25 years ago."

The graphical user interface that came together on the Xerox PARC Alto – or "interim Dynabook," as Kay called it – combined the fruits of earlier research, including icons, windows, and the mouse. But the thing, according to Kay, that brought it all together, that "consolidated these ideas into a powerful theory and long-lived examples" was that his team focused on children. "Early on," he said, "this led to a 90 degree rotation of the purpose of the user interface from 'access to functionality' to 'environment in which users learn by doing.'"

Kay, the visionary, was surrounded by imaginative implementers. He had recruited an educational technology expert, Adele Goldberg, who was also a skilled programmer (she was a co-developer of Smalltalk). She and Kay would come up with research ideas, and they tried out their theories on students in Palo Alto, ages 12 to 15. The programming of different user-interface ideas was often handled by Dan Ingalls, the principal developer of Smalltalk. Many others made significant contributions, including Larry Tesler, who had designed the graphic interface for Charles Simonyi's Bravo. Xerox PARC was an informal setting, where cooperation among people in different research teams was the norm. So people in the computer science laboratory – home to Simonyi, Butler Lampson, Chuck Thacker, and others – talked regularly with people in Kay's learning research group. But the user interface innovation, not surprisingly, came from Kay's team, from a lengthy process of watching how people used the machines and making steady improvements.

Lampson recalled the lack of a user interface on the word processor Bravo, until Tesler came to the rescue. "It wasn't because we didn't think the user interface was important, but we just didn't have the resources in the computer science laboratory," he said. "User interface design is difficult, very iterative work. You can think of things, but you need to try them out on lots of users. . . . The trick is to not let the bits and bytes show through. If they do, you've screwed up in some way."

There were remarkably few bits showing in the Alto system of the late 1970s. So much of the progress in computer science is matter of incremental improvements. Rarely, noted John Shoch, who was at Xerox PARC during those years, are there "wow moments, when you look at something and say to yourself, Aha! But that's what happened with the user interface on the Alto. When people who knew anything about computing looked at it for the first time, they just went, Holy shit!"

Seeing the work at Xerox PARC had that kind of impact on some people at Apple. The shorthand version of the now-famous story has Steve Jobs and a team from Apple visiting Xerox PARC in December 1979, seeing the GUI "fire" there, and taking the torch back to Apple. The facts, as so often, are somewhat more complicated. Well before the visit, the work at Xerox PARC was known and understood by a few people at Apple, notably Jef Raskin, a polymath with degrees in computer science and philosophy, a former visual arts professor and one-time conductor of the San Francisco Chamber Opera Company. Raskin had encouraged the Apple engineers to visit Xerox PARC, and it was Raskin who actually began the Macintosh project in September 1979, before the storied tour.

Yet the visit did alter the thinking of others at Apple, helping to shape the course of its user interface designs for years – first on the Lisa, launched in 1983, and then on the Macintosh that came a year later. One who attended and left impressed was Bill Atkinson, who was the lead designer on the Lisa, along with Larry Tesler, who joined Apple from Xerox PARC. The Lisa, in fact, had many of the graphical interface features later identified with the Macintosh. The two projects overlapped, and often competed for people and resources inside Apple. Indeed, Andy Hertzfeld and others would describe their task in programming the Macintosh as largely "shoehorning" the Lisa user interface onto a smaller, affordable machine.

The Lisa was intended as Apple's big entry into the corporate market, but it was priced at more than $12,000 and arrived nearly two years after IBM had begun selling its far cheaper machine to business customers. Though antediluvian compared to the Lisa, the IBM PC, with its attractive pricing, had a firm foothold in the market, and was good enough for most business people. The Lisa was a technical triumph, but a marketing failure, much as the Xerox Star had been.

Years before the Lisa came out, Jobs had been evicted from the project, with the blessing of adult management that was playing a larger role at Apple. The young Steve Jobs, a volatile perfectionist, given to changing his mind frequently and emotional outbursts, was considered a disruptive force. By 1980, he had his eye on the Macintosh, and he took control of the project the following year, elbowing out Raskin, who left Apple afterwards. The Macintosh that was introduced in 1984 was quite different from the machine Raskin had in mind, which was a simple "information appliance," without even a mouse, selling for about $1,000. Instead, the Macintosh would be more like a modified Lisa, pared down and selling for $2,495. The main champion of that shift was Jobs.

It took a small band of artistic engineers to bring the Macintosh to life. Andy Hertzfeld programmed the working engine of the Macintosh user interface – the software that controlled the screen. His permanent code would dwell in a single, modest ROM (for read-only memory) chip on the machine's logic board. Wedging the look-and-feel of the Macintosh into such a meager memory space was a feat of programming artistry – a kind of haiku in code, his creativity spurred by the hardware limitations of a machine with one-eighth the memory of the Lisa.

To import the best of the Lisa, Hertzfeld would have to husband carefully every bit of memory and every cycle of performance. Most of the Lisa user interface software was written in the Pascal programming language. Instead, Hertzfeld worked closer to the machine, writing in the assembly language tailored to the microprocessor in the Macintosh, the Motorola 68000. Pascal, a higher-level language, required the extra software of a compiler to translate its instructions into the machine code. Hertzfeld did away with that layer of software automation, as if a pilot grabbing the controls himself in tricky conditions instead of relying on automatic pilot. He looked at the output of the Pascal compiler – that is, the translated code from the Lisa – but he could still do far better himself. "I found I could make it between two and three times smaller by hand-coding it in 68000 assembly language," he said.

The radical slimming was accomplished with a grab-bag of programming tricks. For example, Hertzfeld noticed that the Pascal compiler would generate "headers" – statements at the beginning of each programming task – that he could eliminate. The same was true of "trailers" that appeared at the end of each programmed task. He stretched the memory by shuttling certain bits of data into tiny high-speed circuits called registers, instead of holding them in memory. The little beating heart of the Macintosh became more and more robust because of a steady march of such steps. "I could do a better job in assembler since I was working with the actual machine resources instead of an abstraction," he explained. Such hand-coded gains, Hertzfeld observed, would probably be more difficult to achieve today. Compilers have improved and microprocessors have become more complex, executing four instructions at once, for example. "So nowadays the compilers can probably do better than humans in most, but not all, cases," he said.

In part, Hertzfeld regarded his mission as being a technological trustee of Bill Atkinson's work on the Lisa. "It was taking Bill's work and packaging it so that it could reach a market of millions of people instead of thousands," Hertzfeld said. Atkinson was a longtime friend of Guy Tribble, who would become a leader of the Macintosh software team and later joined Hertzfeld at Eazel. Tribble and Atkinson both attended the University of California at San Diego, and they became friends in the computer center. They did a series of computer projects together, including a computer graphics film that took the viewer on an imaginary flight over the human brain. An image of their project, funded by the National Science Foundation, appeared on the cover of *Scientific American*. Both Tribble and Atkinson went to graduate school at the University of Washington – Tribble studied neurology in the medical school, and Atkinson studied biochemistry. To them, computers were mainly tools for the pursuit of science, but their views changed after the microcomputers of the mid-1970s began to appear, suggesting a future for personal computing. Atkinson caught the bug first, joining Apple in 1978, and Tribble followed two years later. Hertzfeld worked alongside Atkinson in his early years at Apple. "I was trained by Bill Atkinson," Hertzfeld recalled two decades later. "He had one rule: make the user happy."

Yet the Macintosh was more than a little Lisa. Its user interface was different in some important ways. The Lisa design revolved entirely around a user opening and closing document files, instead of starting and quitting applications. "It was one hundred percent document-centric," observed Steve Capps, a member of the Macintosh team. On the Lisa, for example, there was no "Apple menu" for launching applications, a fixture of the Macintosh. On the Lisa, the documents were created using the word-processing and other programs supplied by Apple, including LisaWrite, LisaDraw, and LisaProject. The Lisa was a walled enclave – Apple supplied the software, and it was a no-nonsense machine for button-down corporate offices; there were no computer games on the Lisa, for example.

The Macintosh, by contrast, was wide open. It was designed as a launching pad for software programs supplied by outside companies – graphics, games, word processors, any kind of application a developer wanted to make and market. So a user interface for launching applications was crucial for the Macintosh. Depending on outside software companies was risky. Applications appeared slowly and Macintosh sales languished at first; Steve Jobs was a casualty, forced out in 1985. Still, the more open approach was the right strategy, eventually creating a bond of shared dependence and mutual benefit between the Apple Macintosh and hundreds of third-party applications suppliers. The model was to repeat the kind of business ecosystem that sprung up and flourished around the Apple II – a business model, however, that would be truly perfected by the pure software company, Microsoft, rather than Apple.

There were many smaller differences between the Lisa and the Mac. One hotly debated topic, for instance, was the number of clicks required to close a window – two clicks for the Lisa, one for the Macintosh. On the Lisa, the portal into the machine was called Filer, while in the Macintosh it became Finder. Yet more than the name was changed in the Finder, which a person could use to launch programs, to open documents, close them, name them, delete them, and, yes, to find things. In the Lisa, the program was always running on the desktop, ready to go. But the Macintosh did not have the resources for such a luxury. Creating the Finder, the versatile periscope and guide, for the low-power machine was a daunting task. It had to perform its chores, be intuitively easy to use, and not crash – no small accomplishment.

The challenge fell to Bruce Horn, who had been one of the teenage guinea pigs in Alan Kay's Xerox PARC experiments in the 1970s. Horn was struggling, and the shipping deadline of late January 1984 was fast approaching. Steve Capps, who had worked at Xerox before joining Apple, was sent in to help him. "I was just some new batteries, and maybe a fresh perspective," Capps recalled. Horn had been trying to program the Finder according to the Macintosh mantra of trying to make things visual, which meant based on icons. For a while, Capps and Horn tried together to make an icon-based Finder work. But the icons were slowing the execution of the program to a halt – scarcely surprising in a machine with 128 kilobytes of random-access-memory (inexpensive desktop machines in 2001 came with 64 *megabytes* of RAM, 500 times more memory than the 1984 Macintosh).

Capps decided to experiment with a list-based Finder with a list of choices, each a slender ribbon on the screen, dropping down when clicked. He went home and implemented the leaner, less visual version in one night of programming. It ran briskly and, Capps said, "it also felt right for some reason." On the Macintosh team, the "feels right" test was crucial. "But it was also driven by the restrictions," Capps recalled. "There is this myth that has grown up around the Mac that it was some kind of magic. But it was always a compromise. It was time pressure, what was technically possible and what our muse was at the moment. And what really amazes me is that all these years later we are still working in the environment created by choices we made then."

The endurance, and spread, of the desktop metaphor that the Macintosh presented to the world in 1984 has two principal explanations. The first is that the user interface work at Apple during the 1980s was built on research efforts stretching back to the 1960s, whether the self-styled young computing rebels in Cupertino realized it or not. The second explanation is that they really seemed to care about the frustrations of ordinary people trying to use computers, so they generally made thoughtful choices.

The thought process itself is instructive. One of the big debates in the user interface field has been over the number of buttons on a mouse, a debate that continues. At Xerox PARC, the Alto had three buttons, like Doug Engelbart's original mouse at SRI. But in user testing in the run up to bringing out the Star, Xerox noticed that office workers found the three-button mouse confusing. So Xerox put two buttons on the first machine it brought to the marketplace, and Microsoft has followed in the footsteps of the Xerox two-button mouse.

Apple, however, opted for a one-button mouse, first on the Lisa and then the Macintosh. Not everyone likes the one-button mouse, but there was a user philosophy behind Apple's choice. Tribble explained the philosophy with a recollection. He was studying to be a neurologist at the University of Washington when he joined Apple, and while at Apple went back to school to finish his degree. (Even today, Tribble is not only a software engineer, but also a qualified physician taking courses and tests every year to keep his license current.) He remembered being in Seattle, going by a video-game parlor and watching eight- and nine-year-olds play fairly sophisticated warfare and strategy games like Space Invaders and Asteroids. They learned to play not by reading instructions, but by watching an experienced player for a game or two.

It left an impression. "The point was you could watch someone do it and then you could do it yourself," Tribble said. "That was the approach taken at Apple with the Lisa and then more fully realized with the Mac. It was really optimized for ease of learning – ease of use is probably a misnomer. And it's a lot like teaching a new procedure in medical school – see one, do one, teach one. For very visual things, that works. And the Mac was a visual thing . . . And that's why the one-button mouse. With one button, you can watch over someone's shoulder and learn. With more buttons, the clicks are too fine to see."

Nine

Programming for Everyman: Just Let the Users Do It

TRY AS THEY MIGHT, PROGRAMMING LANGUAGES have never really crossed the chasm from the profession to the broad populace of computer users. There have been great strides over the years, from FORTRAN and COBOL to Visual Basic and Java, in opening up programming to an ever-widening circle of professionals. Yet the unrealized promise of the field is that software technology could go further, and allow ordinary users to do the programming themselves. There are hopeful models from other industries. When telephone service first began to expand beyond local communities, it took two or more telephone operators to complete a single long-distance call. There were astronomical projections of the number of operators required for national service. Long-distance telephony looked hopeless, yet with improved message-switching technology and other advances came direct dialing, which automated what had been a labor-intensive job for specialists. Direct dialing, in effect, made everyone an operator.

The programming problem, to be sure, is more complicated than the telephone-operator dilemma of the early 1900s, but do-it-yourself programming is an idea with great appeal. It represents a step beyond the long-running effort to enable more people to use computers – the main aim of user-interface design – to having people increasingly program the machines themselves as well. The payoff would be to spread the benefits of information technology further, to empower users and to boost productivity.

The vision of Everyman and Everywoman a programmer remains elusive, if not an illusion. But there has been progress, which has been achieved mostly with special-purpose tools for specific tasks – database programming, the electronic spreadsheet, and Web site construction. These tools have broadened the programming franchise further, typically being used by a mixture of professional programmers and nonprofessionals. Their success has come from changing the programming perspective from a recipe for *how* to do a certain calculation to a statement of *what* the user wants.

Conventional programming languages are *procedural* in that they are used to describe a procedure for solving some problem – essentially telling the computer, "Do this, then do this, then do this," and so on. But so-called nonstandard languages, such as the SQL database language and the HTML language for creating Web pages, describe what the user wants: summon an assortment of information from a database, or add certain features to a Web site. The hybrid vernaculars like SQL and HTML are thus regarded as *declarative* languages. The modern electronic spreadsheet, created in 1979 by a pair of young MIT graduates, is not a language, but it embraces the same declarative approach as SQL and HTML, focusing on *what* the user wants and then the software interprets the request for the machine. Combining that user perspective, deft design and the computer's tireless capacity for immediate recalculation is what made the electronic spreadsheet not merely a ledger sheet on a screen, but a modeling tool for testing ideas and alternatives.

The search for software that would let the users do the programming began more than 40 years ago. IBM faced the problem at the end of the 1950s. By then, the company had produced several models of computers, and was the nation's leading computer maker. But the market was top-heavy – government agencies, aircraft producers, and other major corporations able to afford the costly machines, and the programmers needed to stir the computers to do useful work. IBM had not yet persuaded its thousands of other business customers to take the plunge. To broaden its base of computing customers, IBM announced the 1401 in October 1959. The machine was priced to appeal to the vast middle swath of business, with monthly rentals beginning at $2,500. The 1401 cost no more than a couple of IBM's large electronic calculating machines and could do the same work, but the 1401 also offered the range of a stored-program computer that could be programmed to do all sorts of tasks.

But the gnawing fear of IBM's business-calculator customers was that acquiring a real computer like the 1401 meant having to hire programmers. Even then, programmers had a reputation as undisciplined free spirits in the work place – a managerial nightmare to be avoided, if possible.

So, in 1959 IBM assigned two of its programmers, Barbara L. Wood and Bernard L. Silkowitz, the task of coming up with a software solution to make the transition to a stored-program computer less wrenching. Their answer was the Report Program Generator, which was introduced in early 1961, a few months after the first 1401 machines were shipped to customers. It quickly became a hit. With RPG – the product was soon known only by its acronym – the user filled out "specification sheets" for a business problem to be computed, such as a payroll calculation. The user specified the input he or she wanted the computer to use, the form of the desired output, and the calculation to be executed. To the user, the RPG method closely resembled the "the keypunch, sort and tabulate mantra" so familiar to veterans of IBM's calculating machines, and its simplified programming "language" required only a few days training. The simple specifications were then translated by software – a generator program – into machine language. RPG was the software bridge from accounting machines to computers. IBM's official historians wrote in 1986 that the RPG concept would "acquire even greater importance in later decades, when further technological improvements brought the computer to a much larger community of users."

Don Chamberlin grew up in Campbell, California, when it was a little market town near San Jose, known for its prunes and apricots, before there was any reason for the term "Silicon Valley." He was a member of the "Sputnik generation," an eighth grader when the tiny Soviet satellite was launched, a time when American schoolchildren were encouraged to explore technical fields. Chamberlin, the son of a high-school principal, did not require much encouragement. He was good at science, and loved to build things – model planes, trains, and rockets. The family surrendered part of its modest garage to his various kit-building projects. Chamberlin first encountered computers at Harvey Mudd College, a small elite California school specializing in math, science, and engineering. In the early 1960s, he recalled, "Computers were still big machines in the basement, but there was this sense that they were the future." There were sign-up sheets to get a half hour on the IBM mainframe, and one of his first programs was a tic-tac-toe game, written in FORTRAN, just for fun.

An excellent student, Chamberlin got a National Science Foundation scholarship to do graduate work at Stanford University and, after earning his Ph.D., he joined IBM in 1971 as a researcher at the company's Watson Labs in suburban New York. His first project, an operating system for an experimental timesharing computer, folded within a year, and his boss decided that the group should focus instead on database software. In the early 1970s, database technology was a lively field of research, fueled by advances in data storage and accelerating demand from corporations eager to use computers more effectively. In the early days of computing, data were stored sequentially on tapes, each of which handled a particular job, such as inventory control, purchasing, or accounts receivable. The data were trapped on the linear tape, making it difficult for a company to use its business data for more than one application. That changed, starting in 1956, when IBM developed the first magnetic hard disk. The hard disk allowed for "random access" to data anywhere on the disk. Data were no longer hostage to the linear sequence of a tape. They could be organized in all sorts of ways, spread across the disk as if trees of data, graphs, or networks. The hard disk really opened the way to innovation in database software – programs that could access, manipulate, and organize information. The potential became more and more apparent by the early 1970s, as disk drives improved along the same trajectory as microprocessor chips, doubling the density of data that can be stored on a square inch of disk every 18 months and drastically reducing data-access times and cost – a march of progress that continues apace. (In 1956, it cost $10,000 for a megabyte of disk storage, or 1 million bytes of data. In 2001, it cost less than $1,000 for entire personal computer, which included a hard disk with a capacity of 20 gigabytes, or 20 *billion* bytes of data.)

When Chamberlin began his research in the early 1970s, the database field was shaped by the work of Charles Bachman, a software designer at General Electric and later Honeywell. In his 1973 paper, "The Programmer as Navigator," Bachman described database technology as a tool for navigating through stored data records using software pointers as guides, as if following a path in a forest. Chamberlin became schooled in the intricate, navigational approach. "I just loved it and thought it was neat," he recalled. "You could study it all day. It was a real puzzle." Chamberlin and others at the IBM lab in Yorktown Heights, New York, were vaguely aware of some research being done in the company's San Jose lab by an Oxford-educated mathematician, E. F. (Ted) Codd. At the time, Chamberlin remembered, his initial impression of Codd's work was that it looked odd. "Some kind of strange mathematical notion. Nobody took it very seriously."

Yet Chamberlin took it seriously indeed after Codd visited Yorktown Heights in 1972 and gave a lecture on his work. Codd explained his "relational" model for databases. Simply put, Codd saw data as things, and the relations between things – for example, employees, salaries, office locations, and the like. The related data could be grouped in tables. Data, Codd insisted, should not be viewed as a forest-like network that had to be navigated. Instead, he argued, data should be stored in tables and then linked in various ways. In the relational approach, the user should be able to state a query, or a request for data, in a declarative way, much as a person might ask a question. At the 1972 lecture, Chamberlin recalled Codd stating hypothetical queries like, "Find employees who earn more than their managers?" Sitting in the audience, Chamberlin would imagine what such queries would look like if he used the navigational method – "five pages long that would navigate through this labyrinth of pointers and stuff," while Codd, he recalled, would "sort of write them down as one-liners." To Chamberlin, it was "a conversion experience," the moment when he really grasped the power of Codd's idea.

Codd's ideas, in fact, had been around for some time by then. His paper, "A Relational Model of Data for Large Shared Data Banks," was published in 1970. In it, Codd declared that the network, navigational model was seriously flawed, having "spawned a number of confusions." His relational model, he insisted, was "superior in several respects." So it would prove to be, but at the time Codd's idea was not only a contrarian assault on the prevailing wisdom of database computing, it was also a theoretical proposal rendered in an arcane mathematical notion. Most people in computing did not understand what Codd was talking about, but Chamberlin did that day in 1972, and he also had his first inklings of how to convert the theory into working database software. The IBM project would be called System R, and the database programming language created mainly by Chamberlin became SQL, for Structured Query Langauge.

In 1973, Chamberlin and a few others from Yorktown Heights moved out to San Jose to work on System R, the relational database prototype. Though a research project, the System R team was essentially a cutting-edge engineering task force. Their mission was to implement theory, not ponder it. They built a working system, from industrial-strength transaction-processing software to a specialized language intended for non-experts. Codd himself would not be part of the System R team; he was more a visionary than a project leader or manager. The research chiefs back East considered giving Codd a role in System R, but they had no intention of letting him run it. So Codd kept his distance as an occasional consultant to the project, but mostly he preached the relational database gospel at academic conferences and to IBM customers, and moved onto other research interests. Chamberlin described Codd's role by analogy. "Ted Codd provided the concept for a new house, and the people who designed and built System R were the ones who prepared the detailed blueprints, poured the foundation, and did the carpentry work. Ted by his training and orientation was a theoretician rather than a system builder. A gifted architect is not always a very effective carpenter."

The System R team, which eventually included 20 people, was mainly a group of young researchers, led by a few seasoned managers. It was a new field, so there were no established "experts," certainly not in relational databases. The young researchers came from varied backgrounds, and for different reasons. Jim Gray was working at Yorktown Heights on research that applied computing to social problems; IBM did a traffic flow optimization model for New York's Holland Tunnel, for example. It was a time when there was great optimism that computer simulations could be used to great benefit by social planners. The work of Jay Forrester of MIT had been used by the Club of Rome as a basis for *The Limits To Growth,* a dystopian warning in 1972 of coming resource shortages and population explosion unless sweeping government programs were adopted. One of Gray's projects was to show Forrester's models were flawed.

The work was fascinating, but Gray, a native Californian, hated living in the East, and he particularly hated its dark, cold winters. He told his boss he didn't think he could bear another winter in New York, and asked for a transfer to California. There were no openings, he was told, since IBM had a freeze on hiring in the midst of an economic downturn. Besides, his boss thought Gray was overstating his dislike of snow and slush. Gray, however, was not kidding, and he was determined to flee suburban New York. "California," he recalled, "seemed like paradise by comparison, warm sun and warm people and ripe tomatoes all year round." Gray quit his job, hopped in his Volkswagen Beetle, and drove across the country back to California. He showed up at the doorstep of the IBM labs in San Jose, which found an opening for him. Gray joined the relational database group and went on to define the basic properties for transaction-processing software – the vital underlying technology that takes the information in databases and then deploys it to animate the world's financial markets, airline reservations systems, and bank automatic-teller-machine networks. In 1999, Gray, who became a researcher for Microsoft, won a prestigious Turing award for his work, which began after his serendipitous flight west.

Once underway, the System R project had the enthusiastic support of Ralph Gomery, the head of IBM's research division. The System R work touched on a Gomery passion – making computers easier to use. And database software was increasingly becoming a big business. "This was not unfettered research," Gray reflected. "This was research focused on revolutionizing the way we access and organize information."

In his writing, Codd had proposed two possible languages for relational databases, both deeply mathematical with lots of Greek letters and subscripts (numbers and symbols below the line of type). Codd's proposed languages could be read only by trained mathematicians, and could not be typed in on a keyboard. "Clearly, this wasn't going to be a big commercial hit if Ted's idea depended on his notation," Chamberlin said. So Chamberlin and a colleague, Ray Boyce, began to tackle the problem of the user interface for the new database system. Chamberlin and Boyce collaborated closely in the early days on SQL, but Boyce died tragically in 1974, a year into the project, of a brain aneurysm. Their goal was to devise a special-purpose language that could be understood easily by humans and compiled by the machine. The relational approach made it possible to ask questions of a database in a way that was more natural to humans, because the questions themselves did not have to include the path to the data as they did in the navigational model.

To make it work, however, the language had to be designed and planned precisely, at just the right level, with enough structure so the compiler could translate it for the machine and enough natural-language familiarity to make it easy for humans to use. The language that Chamberlin and Boyce designed resembles a pidgin English. In fact, it was originally called SEQUEL (Structured English Query Language), until the IBM lawyers discovered that SEQUEL was a registered trademark held by a British aircraft company. Just as well, Chamberlin says, since the reference to English is a bit misleading. "It wasn't English because English is a terrible language for data queries," he explained. "It's very ambiguous."

To give SQL a regular structure that could be processed by the computer, the "verbs," or data manipulation commands, were limited to about eight basic concepts, such as SELECT, FROM, WHERE, GROUP BY and ORDER BY. The data to be queried were also stored in a structured format – the tables of related data that Ted Codd described in his paper. These tables might include employees (EMP, for short) with their names, departments, locations, jobs, and salaries in a big table. So the query, "Find Mary Smith's salary," is rendered in SQL as:

```
SELECT          SALARY
FROM            EMP
WHERE           NAME = "MARY SMITH"
```

In another, somewhat more complicated example, a business may want to examine the credit limits of certain customers. The customer data would be stored in a relational database and might include the name, city, state, zip code, and credit limit for each customer. So the query, "Find customers with a credit limit of more than $10,000 and order them from highest to lowest credit limit," is rendered in SQL as:

```
SELECT          NAME, CITY, STATE, ZIPCODE
FROM            CUSTOMER
WHERE           CREDITLIMIT > 10000
ORDER BY        CREDITLIMIT          DESC
```

Today, more than 90 percent of the world's business data are stored in relational databases, and Chamberlin's SQL is the dominant database language. That happened because of the work of many others, including Pat Selinger, who wrote the "optimizer" that translated SQL into a detailed plan for retrieving data and Jim Gray, who fashioned the transaction-processing technology that made SQL so useful in modern electronic commerce. Yet Chamberlin is the one who synthesized things, who saw how the pieces would fit together – the data definitions, the data manipulations, and the data tables. He saw the importance of Ted Codd's ideas, and a way to make them work. Alan Kay once said, "Point of view is worth 80 IQ points." Chamberlin brought a distilling and simplifying perspective to the field of relational database software. "Don was enough of a programmer to see how to translate these ideas into reality," observed Jim Gray. "All of this seemed obvious to Don, and when he showed it to us, it was obvious to us too."

IBM was slow to bring SQL to the marketplace. In 1979, the first SQL data-base product was introduced by a Silicon Valley startup called Relational Soft-ware, Inc., which changed its name to Oracle Corporation three years later. That happened partly because IBM's labs published its research. Others work-ing on relational databases, such as Berkeley's government-funded Ingres proj-ect led by Michael Stonebraker and Eugene Wong, published their research as well. So the ideas being developed about relational databases were in the air, waiting to be exploited by enterprising entrepreneurs like Lawrence J. Ellison, founder and chief executive of Oracle, who would become a billionaire many times over. (Displaying an early flair for imaginative marketing, Ellison called his first product Oracle Version 2.0. There was no version one, since knowl-edgeable customers avoid Version 1.0 of most software, figuring it's riddled with bugs.)

In the summer of 1978, Ellison phoned Chamberlin and told him how use-ful he had found a paper published in 1974, "SEQUEL: A Structured English Query Language," written by Chamberlin and Boyce. "We wrote the paper that Larry Ellison based his company on," Chamberlin observed. Ellison was not merely making a social call. He was interested in obtaining detailed technical information – called "error code values" – from Chamberlin so his product would be fully compatible with IBM's work. IBM decided that there were lim-its to how much it would help a rival, even a startup, and declined. IBM's cor-porate management was reluctant to adopt the SQL relational database tech-nology, because in the late 1970s its IMS navigational product was thriving. Still, it seems an irony that it was outside competition from Oracle and others that prodded IBM to take seriously the remarkable database research that was being done in its own labs.

With its use of familiar terminology and its focus on human questions instead of navigating the underlying bits of a computer, SQL has made it far easier to build database applications. People use SQL systems every day without knowing it. When a credit-card strip is read by a store machine, or when a person punches numbers on an ATM machine, those numbers are being used by SQL statements to access a database and update the person's credit or bank account. For all its success, however, SQL fell short of one of its original goals. Initially, Chamberlin and his colleagues thought SQL would make it pos-sible for ordinary people to interact with data in new ways by using SQL them-selves. For the most part, it did not turn out that way. The person at the ATM is accessing a huge database directly, but the SQL is hidden behind buttons or a touch screen.

For Chamberlin, the realization that SQL would have limited appeal beyond professional programmers came in 1975. The System R project recruited students from San Jose State University to see how a random sample of college students reacted to SQL. The results were not encouraging. With enough hard work, they mastered it, but it was slow going and the students made a lot of mistakes. "We were too optimistic in 1975 about the use of computer languages by the general public," Chamberlin observed one day in his IBM office, overlooking the dun hills outside San Jose. "Computers take every instruction very literally and have no common sense at all. So even though a language like SQL has English words, it does not have the colloquial flexibility of a natural language."

Dan Bricklin and Bob Frankston met in 1970 at MIT. Bricklin came from Philadelphia and Frankston from New York, but their backgrounds were remarkably similar. The pair would go on to develop a product that would help jump-start the personal-computer revolution in 1979, and they remain friends, living a short drive from each other outside Boston. When asked why the two hit it off so well from the outset, Frankston, sitting in his home office in Cambridge – a personal-computer lab crowded with machines, gadgets and parts – replied, "It's always hard to say why, but it just made sense."

Both came from came from upwardly striving Jewish families, where intellectual pursuits were encouraged and entrepreneurial gumption put food on the table. Bricklin and Frankston would both display a fascination with computers, and they would find a way as teenagers to get their hands on the big institutional machines of the 1960s. Bricklin's father, Baruch, ran the family printing business in Philadelphia, but he had a mechanical engineering degree from the University of Pennsylvania. "Dad used to put me on his knee, and teach me how to use a slide rule," Bricklin recalled. "An engineering bent definitely ran in the family." As a child, Bricklin soon graduated from Erector sets to electronics, using Heathkit sets to build a shortwave radio, a stereo that the family used for years, even a kitchen intercom so his mother Ruth could summon him to meals from his upstairs bedroom.

Bricklin did his first programming at 15 on a high school time-sharing terminal, and he began by writing some simple programs that graphed numbers and sorted text. In 1967, Bricklin earned a place in a summer program for high-school science students that allowed him to access the computers at the University of Pennsylvania's Moore School of Engineering, working just down the hall from the room where the original ENIAC machine was built. He then made himself useful at Penn's Wharton business school by creating a program used to grade standardized tests automatically. His program for adding some specialized extensions to FORTRAN, which he called WhartFor, later won a national science prize for high-school students. In the fall of 1969, Bricklin entered MIT intending to major in mathematics, though he soon abandoned math for computer science.

Frankston's father, Benjamin, ran a small manufacturer of electronic parts that he sold to television repair firms. His mother, Dorothy, taught geology and geography at Hunter College. At his mother's insistence, the family name was changed to Frankston from Frankenstein. Though pronounced Frankenshteen, the name had its drawbacks, regardless of the pronunciation, given the popularity of Mary Shelley's Gothic horror story and its many film versions. "You can imagine my uncle Boris trying to sell life insurance," Frankston said, smiling and shaking his head. At 13, he wormed his way into evening classes on how to use computers at Hunter, where his mother was a teacher. The classes were for the faculty, but no one seemed to object to the teenager in their midst. And no Hunter faculty member would ever make as good use of the instruction as the skinny kid in the back to the classroom. His first program, in FORTRAN, printed out the leap years for decades and decades.

At Stuyvesant High School, a highly selective public school in New York City, Frankston was a bright but distracted student, impatient with the course work. One of the advantages of attending Stuyvesant was that students could get access to the IBM mainframe at nearby New York University. In programming, Frankston says he found an activity that "just happened to resonate with me." He enjoyed the technical virtuosity required to get a program to work, enjoyed the larger goal of building something, and enjoyed the tactile sensation of interacting with the machine. "It's a drug," he said. As he got better, Frankston found that there was a waiting market for his skill. He got summer jobs and part-time work as a programmer. In his senior year, he helped teach Stuyvesant's programming course – "the only class in high school I ever got a 100 in," he recalled. One part-time job was at a Wall Street investment bank, White Weld & Company, working on database and calculation software used to analyze trends in the stock and bond markets. Even at MIT, Frankston held onto the

White Weld job, commuting to New York occasionally and working from a Teletype machine in his apartment, running up phone bills of $1,500 a year. "You had to have a job to really learn anything," he explained.

Bricklin and Frankston met in 1970 as programmers on MIT's Project MAC, which had embarked on its ambitious Multics time-sharing system. The Multics program – a joint effort by MIT, General Electric, and AT&T's Bell Labs – employed roughly 100 software developers, but only a handful of students like Bricklin and Frankston. They would typically show up in the evenings and work all night. The two quickly became friends, often driving off at dawn in Frankston's car to have breakfast together. Even then, they talked of one day going into business together. "He and I were entrepreneurial types," Bricklin recalled.

Yet it would be nearly nine years before Bricklin and Frankston founded Software Arts to make a business of their electronic spreadsheet program, called VisiCalc. At MIT, Bricklin wrote a simple calculator program for Multics, and worked on compiler and interpreter programs. After graduating, he got a job in Digital Equipment's typesetting group. The job had Bricklin traveling to newspaper offices in the United States and Canada that were making the transition from craft-printing methods to computerized typesetting and screen-editing systems. He vividly recalled working round-the-clock at the *Kansas City Star* in 1974 to debug the electronically-transmitted transcripts of the Nixon Watergate tapes. At Digital, he had to design word-processing and editing software for people who, if anything, disliked computers – a world away from the MIT computer lab and the Multics project. "It was an education in user-interface design," he recalled, because "newspaper people are real users." In 1976, Bricklin took a programming job at a producer of computerized cash registers. He was working to save money for Harvard Business School, which he would attend the following year. But the job also gave Bricklin a taste of close-to-the-metal programming for the Motorola 6800 microprocessor, an ancestor of the microchips that power Apple personal computers today.

For his part, Frankston, who is two years older than Bricklin, stayed on doing graduate work at MIT until 1976. At MIT, Frankston was a founder of the Student Information Processing Board, a time-sharing project intended to give university students access to computing. His master's project focused on how computer networks might become a marketplace for services – an early vision of electronic commerce and online transactions. All the while, he continued to do contract programming work for companies, including White Weld, until 1979.

In broad strokes, the story of VisiCalc told most often is that a Harvard Business School student and his buddy, outsiders to computing, invented the electronic spreadsheet, the first "killer app" of the personal computer industry. That version is true, up to a point. Yet VisiCalc was created by two young men in their late twenties, who had deep and varied experience in computing – in databases, financial programs, compilers, interpreters, text editing, and user-interface design. "All our computer science background went into the mix that became VisiCalc," Frankston observed. "So much for this myth about a couple of guys coming up with something out of nowhere."

In the fall of 1977, Bricklin entered Harvard Business School. By the next spring, he was sitting in Aldrich Hall, Room 108, daydreaming about a better way to do all the financial calculations. For figuring a company's sales and profits by month, quarter, and year, factoring in variables that ranged from marketing and manufacturing to shipping and interest charges, the routine was the same. It involved separate calculations, written on sheets of graph paper, added, subtracted, or multiplied as required. When assumptions were changed, new calculations had to be made, each one a fresh chance for error. Such labor, of course, had been the lot of business-school students, managers, and financial analysts for generations. Handheld calculators such as those made by Texas Instruments and Hewlett-Packard made it less tedious than in the past, but Bricklin figured there had to be a smarter way to do it. Computers, after all, were made for doing constant streams of calculation.

One of the amenities at Harvard was free computing on a time-sharing system that used the BASIC programming language. Bricklin tried it, but the delays in the time-shared system running BASIC made it too slow for the kind of rapid calculation and recalculation he wanted to do. Bricklin had seen a Xerox PARC Alto machine at MIT, appreciating both the quick interactive responsiveness of the personal machine and the visual richness of its graphic text-editor, Bravo. Yet the Xerox lab in Palo Alto never tried a spreadsheet. "We had no accountants at PARC," Butler Lampson explained.

Bricklin may not have been an accountant, but both he and Frankston had studied accounting, and both of them, unlike most researchers at Xerox PARC, were actually interested in business. So Bricklin addressed the challenge of designing an electronic spreadsheet. He thought about the issues of how numbers and text would be presented on the screen, how calculations would be done, and how to handle the formatting and data structures so they would be understood by the user and properly interpreted by the computer. Bricklin strove for more than merely the automated equivalent of a paper spreadsheet. He wanted flexibility – the ability to present the user different perspectives – through the magic of programming. "I felt you should be able to switch views," he said.

Bricklin decided to run his idea by some free consultants – his Harvard professors. At that point, all he had was an idea. The first couple of professors he spoke to were encouraging, but his finance professor told him there were a lot of FORTRAN financial programs already on the market. He told Bricklin to talk to a second-year business school student, Dan Fylstra, who was dabbling in the computer business himself with a startup called Personal Software. Fylstra, Bricklin recalled the professor saying, would be able to confirm that Bricklin's idea was misguided. Bricklin called Fylstra, talked to him briefly, but did not tell him what he was working on.

Bricklin kept thinking about his spreadsheet idea over the summer of 1978, and he did a little research of his own. Yes, there were financial forecasting programs on the market, but they were mainly for corporations, often costing thousands of dollars a month on time-sharing systems. In the fall, Bricklin got in touch with Fylstra again, discussed his idea more specifically, and Fylstra loaned him an Apple II to hack together a "weekend prototype." The crude mockup had columns and rows, and did some simple arithmetic but no more. Fylstra, who wanted a checkbook program to market, encouraged Bricklin to go ahead. Bricklin called his old friend, Bob Frankston, and asked him if he wanted to go into business together, as they had discussed so often in the past. They soon had a plan and a division of labor. Bricklin would stay in school at Harvard, but he would do the design work and Frankston would be the programmer. "Bob went off and built the thing," Bricklin recalled.

They worked in a style that Frankston described as "fast-iteration prototyping" – two people working together closely on a program that evolved through constant modification and experimentation. A crucial design decision was the use of the "grid," resembling the grid pattern on paper used by accountants and engineers. "You have to give people a metaphor they can grasp," Bricklin noted. The grid was more than a familiar appearance; to the computer, the grid was a data structure, each cell in the grid being a computer location for data. The user could change numbers, move them to different cells in different order, and see that the computer had them, ready for instant calculation, by columns, rows, or selectively. It gave VisiCalc a flexibility that the more linear financial forecasting programs did not have. It was a visual tool − a fact emphasized by the product's name, a contraction of "Visible Calculator" − and a powerful one, largely because of the personal computer's capacity for immediate recalculation. VisiCalc was a means for asking all sorts of what-if questions. "You didn't have to plan out the program in advance," Frankston explained. "You worked with it as if modeling clay. It was a tool for working out your ideas."

For its time, VisiCalc was a tight, fast program. The user could scroll smoothly up and down, left and right. Cursor movements and keyboard strokes registered instantly on the screen. It had a peppy, interactive feel. That was no mean feat, considering that Frankston programmed VisiCalc for the Apple II, which was designed as a game machine. Though infinitesimal by modern standards, VisiCalc used every bit of memory and processing power the Apple II could muster; VisiCalc took over the machine. The Apple II had no operating system, so Frankston, writing in assembly language, had to program the operations down to the tiniest detail. If, for example, a person typed too fast Frankston had to write the instruction that would "poll" that keystroke − momentarily store the character and then present it on the screen an instant later, ensuring the keystroke was not lost or appeared only after an irritatingly long delay.

VisiCalc was a masterful work of design, detail, and construction with minimal means. Not everyone, but some of shrewdest minds in the industry, recognized VisiCalc as a breakthrough. The most eloquent was Benjamin Rosen, an analyst at the investment bank Morgan Stanley & Company. Later, Rosen would go on to become a venture capitalist and one of the principal backers of Lotus Development and Compaq Computer; for years, he served as Compaq's chairman. Months before VisiCalc went on sale, Rosen was shown the electronic spreadsheet.

In a report to the investment bank's clients on July 11, 1979, Rosen wrote, "In mainframes and minicomputers, hardware developments have always outpaced those of software. This pattern is now being repeated in the nascent field of personal computers. Today, virtually the only user of personal computers who is satisfied with the state of the software art is the hobbyist. And he does all his programming himself. . . . Though hard to describe in words, VisiCalc comes alive visually. In minutes, people who have never used a computer are writing and using programs. Although you are operating in plain English, the program is being executed in machine language. But as far as you're concerned, the entire procedure is software transparent. You simply write on this electronic blackboard what you would like it to do – and it does it."

"VisiCalc," Rosen concluded, "could someday become the software tail that wags (and sells) the personal computer dog." VisiCalc played precisely that role. It was the demonstration product that showed the personal computer was not merely a hobbyist enthusiasm. It could be used as a business tool, among other things. Many people bought Apple II machines solely to run VisiCalc. As it grew, Software Arts ran into troubles. It was slow to develop VisiCalc for the IBM PC, which quickly became the mainstream personal computer in the business market. Bricklin and Frankston quarreled with their publisher and marketer, Dan Fylstra, and lawsuits followed. Software Arts sold out to Lotus Development, which developed the leading electronic spreadsheet for the IBM PC, Lotus 1–2–3. Later, Lotus was supplanted by Microsoft and its Excel spreadsheet.

Bricklin and Frankston did not become rich from their creation, yet Visi-Calc was the canonical program; it showed the way, and others followed in its path. In programming terms, VisiCalc took the declarative programming of RPG and SQL a step further so that, at least in the field of financial modeling, ordinary people could program. They did not have to write algorithms specifying *how* to do something; they could describe what they wanted done, and VisiCalc would do it. The method has been called "programming by example." At a 1995 symposium reuniting the team that worked on SQL, a former IBM researcher mentioned that, during the late 1970s, there was a group at IBM's Yorktown Heights lab working on programming by example. "If you showed a fancy piece of software what you wanted as the output, it would figure out how to get it," explained Mike Blasgen, the former IBM researcher. "That never came to anything. Actually, do you know what programming by example is? It really actually happened. It's VisiCalc."

"If you say 'HTML programming,' every real engineer snickers. HTML isn't considered programming," observed Brian Behlendorf, a leader of the collaborative Apache project, whose popular software helps power the Web. "But the reason there are billions of Web pages is that making them in HTML is easy to do."

Hypertext mark-up language is the lingua franca of the World Wide Web, a basic vernacular of document presentation and interchange over the Internet. And HTML, along with its acronymic cousins URL (uniform resource locator) and HTTP (hypertext transfer protocol), are the three technical pillars of the Web – a trio of disarmingly simple yet powerful software standards behind a new global medium of information and commerce. The Web has had as great an impact as anything in computing over the last 25 years or so and, like most computer advances, it was enthusiastically embraced and commercially exploited first in the United States. Yet the Web was not the work of a computer scientist at a leading American university or corporation. It was, instead, the creation of a British physicist, working in the shadow of the Alps at the CERN physics laboratory outside Geneva.

Tim Berners-Lee, as it turned out, was ideally trained for the problem he addressed more than a decade ago when he fashioned the Web at the European research center for particle physics. He may have even had a genetic advantage, being the son of two English mathematicians, Conway Berners-Lee and Mary Lee, who were computing pioneers in their own right. They were members of the team that programmed one of the world's first commercial stored-program computers, the Ferranti Mark I in 1951. As a teenager, Berners-Lee recalled his father working on a speech for Basil de Ferranti, the chairman of the electronics company that bore his name. To prepare, Berners-Lee's father was reading about the brain, searching for ideas on how computers – literal-minded number crunchers – might be adapted to make the kind of intuitive connections that humans do. It struck the boy as an intriguing notion, and it stayed with him.

In 1976, Berners-Lee graduated from Queen's College at Oxford University. His degree was in physics, but he was an engineer by instinct. Fascinated by computing and the arrival of affordable microprocessors, Berners-Lee built his own personal computer, cobbling together a Motorola 6800 microprocessor, a jury-built motherboard, and an old television for a monitor. He went into software engineering and design, and his early career was a combination of work at big companies, consulting jobs, and contract assignments. He was first hired by Plessey Telecommunications, a big telecommunications equipment supplier in Britain, where he worked on projects ranging from message relays to bar-code technology.

Berners-Lee then joined a manufacturer in Dorset, D. G. Nash Ltd., where he wrote typesetting software for printers. Afterwards, he became an independent contractor for about a year and a half, which included a six-month stint in 1980 at CERN, where he put together a program for himself for storing personal information he called "Enquire," which had some linking features, a conceptual forerunner to his later work. He next worked on a friend's startup, applying microprocessor technology to inexpensive dot-matrix printers to enable them to print rich graphic images. On that project, he wrote a mark-up language for documents handled by the printer.

By 1984, when he returned to CERN, Berners-Lee had eight years of varied, even scattered, experience in computing. Much of his time, though, had been spent dealing with the practical problems of handling documents. Throughout those years, he had continued to ponder the larger issue of how to use computers to organize information more efficiently. "Looking back," Berners-Lee recalled, "I had a career that moved around a bit that enabled me to pick up the skills I needed to design the Web." Back at CERN, he set to work on software and information-management issues. There he encountered a sprawling institutional environment of the pre-Internet days: thousands of researchers working on computing systems from many vendors, each a separate island of information. It struck Berners-Lee as terribly inefficient, and the problem was magnified, he said, by the "irritating habit that physicists had of writing all their software themselves." What was needed, Berners-Lee decided, was a networked version of his "Enquire" program for storing and linking information – a tool for more productive, collaborative work.

Over the next several years, the European particle physics lab became Berners-Lee's incubator for the Web. He set out simply to create a better working environment at CERN, but the solution he eventually devised was a general one, useful to many others. The physics lab had no real interest in computer science research as such. Berners-Lee's project had the immediate objective of using databases and documents more efficiently at CERN, though he was always thinking broadly. And CERN provided a freewheeling, well-funded setting for Berner-Lee's wider explorations into the realms of hypertext and the Internet – two technologies that Berners-Lee would imaginatively join with software.

The concept of a vast automated storehouse of knowledge goes back at least to Vannevar Bush's "As We May Think" article in the *The Atlantic Monthly* in 1945, in which he spoke of a personal information machine called a memex that could store millions of books, newspapers, magazines, and other published reports. Douglas Engelbart's work in the 1960s at Stanford Research Institute centered on making on-line information accessible in order to "augment human intelligence." Ted Nelson, the iconoclastic evangelist for power-to-the-people "computer liberation" that influenced thinking in the personal computer community, coined the term "hypertext" in the mid-1960s – a system for navigating through a world of information stored on computers. As he pursued his project further in the late 1980s, Berners-Lee studied the "prior art" of the field. A European colleague who had worked in the United States, seeing first-hand how universities and research labs across America were linked by the Internet, became an advocate for its adoption at CERN. At the time, there was scant interest in Europe in the Internet, but Berners-Lee was impressed. The Internet's democratizing connective power, he recognized, could amplify his notion of a shared information network. It could become "global hypertext," it could become the World Wide Web.

Berners-Lee made the rounds of the hypertext and electronic book communities, hoping to convince them to adapt their products for the Internet. That way, he figured, he could simply buy the tools he needed, instead of having to build them himself. That was not an unreasonable expectation. In 1987, for example, Apple Computer had introduced HyperCard, an easy-to-use system of information cards that were connected through hypertext-style links. A user could create information on cards – documents, addresses, schedules, timelines – and then click to jump from one card to another. But the hypertext and electronic-book businesses that Berners-Lee spoke with did not grasp the potential impact of the Internet. At the time, the Internet was the bailiwick of a comparatively small group of universities and researchers. Its language was Unix, and its culture was geeky. Nobody, including Berners-Lee, could have then predicted the Internet explosion that would come in the mid-1990s, fueled by the astronomical growth of the Web and by the point-and-click browsing software, first commercialized by the Netscape Communications Corporation that made it easy to navigate the Web.

So Berners-Lee went ahead on his own. By the end of 1990, he had fashioned the prototype versions of the URL, HTTP and HTML – the fundamental standards required for addressing, linking, and transferring multimedia documents. He built them using the hardware and software of the NeXT Inc., the work-station maker founded in 1985 by Steve Jobs, after he was pushed out of Apple. Even years later, Berners-Lee would sing the praises of NeXT, its implementation of the Unix operating system, and its user interface – "beautiful, smooth and consistent" – and its software for creating hypertext programs. In 1991, just as he released his software over the Internet, Berners-Lee visited Jobs and Guy Tribble, who had left Apple to join Jobs as NeXT's head of software engineering. But the NeXT team did not grasp the significance of Berners-Lee's work either. "It's really hard to demo the World Wide Web if there's no World Wide Web yet," Tribble recalled. "There we were at NeXT, scratching our heads. After he was done, we all kind of said to each other, 'Well, we don't see it.' We missed it like everyone else at the time."

Three years later, with the Web on the verge of becoming a mainstream medium of information and commerce, Berners-Lee moved to the MIT Laboratory for Computer Science, where he became director of the World Wide Web Consortium, the organization that sets standards for the Web. His mission is to keep the Web a universal information space, open to all, rather than being balkanized into separate commercial camps. A slender man with an unruly head of thinning blond hair, Berners-Lee has been an oddly compelling authority figure so far, part pragmatist and part Internet idealist. He is no hair-shirt purist. In the early 1990s, he said, thoughts of how to cash in on his creation "crossed my mind almost every day." In the end, however, working independently to keep the Web open and outside the influence of any one company seemed like the right thing to do.

Berners-Lee speaks as fast as he thinks – sometimes faster. "Now where was I?" he inquires, breaking off in mid-sentence. His multilingual colleagues at CERN eventually asked him to speak in French, hoping to slow him down. He has the English touch with self-deprecating humor. In front of an audience of 1,000 people or so at a Web conference, he encounters the inevitable computer foul-ups in demonstrating some software. Puttering at the keyboard as he fixes the problem, he says as an aside, "Oh, you'd all be talking amongst yourselves if we set this up properly."

According to Berners-Lee, the Internet is a universe that embraces the social values of "tolerance and decentralized control" that he echoes personally. "Why is censorship wrong?" he asks rhetorically. "Because it is a centralized notion of what is good," suggesting a rational engineering perspective instead of a subjective moral judgment. The value system of the Internet, he says, translates into the software design principles of "simplicity and modularity."

In designing the Web, Berners-Lee was guided by those values and design principles. The uniform resource locator, or URL, originally called the universal resource indentifier (URI), can "point to any document (or any other type of resource) in the universe of information" – a software standard as tolerant and decentralized as one could imagine. The hypertext transfer protocol, or HTTP, is a communications standard for sending a Web page from a data-serving computer to countless desktop computers, or even handheld devices. Hypertext mark-up language is a simple way to represent hypertext, so it can be broken up, sent over the Internet and reassembled as multimedia Web pages on millions of desktop screens each day.

Computer scientists who looked at Berners-Lee's creation at first thought it not merely simple, but simple-minded. No real computer systems engineer, they said, would ever design something like the Web, in which every click opens a new Internet transmission protocol connection. Too messy and chaotic, some systems researchers said, who predicted the Web would not be able to handle ever-increasing volumes of use. But the Web has probably scaled up better than any software artifact in history. Berners-Lee made some shrewd decisions on the HTML language. He used the standard syntax of the existing family of mark-up languages – the standard generalized mark-up language, or SGML, familiar to professional document specialists. Bowing to SGML made it easier to gain acceptance for HTML in the hypertext community, which was important initially. It is the SGML legacy that gives HTML its trademark instructions, called "tags," which are formatting commands between angle brackets. So <title> is the tag that begins a title, <body> begins the body of the document, <b> begins bold text, <u> for underline text, <img src> for insert an image. The initial version of HTML contained only such basics, while later versions added automated features, such as style sheets that contain a package of layout settings.

Like RPG, SQL, and VisiCalc, HTML is a tool for a simplified declarative style of programming – the *what* description of a problem or a Web page, rather than the procedural, closer-to-the-machine *how* of traditional programming. In

Berners-Lee's case, he did not initially think of HTML programming as something that nonprofessionals would do, but there was no stopping them once the Web took off. In his book *Weaving the Web* (written with Mark Fischetti), Berners-Lee wrote, "I never intended HTML source code (the stuff with the angle brackets) to be seen by users. . . . But the human readability of HTML was an unexpected boon. To my surprise, people quickly became familiar with the tags and started writing their own HTML documents directly."

10

Java: The Messy Birth of a New Language

JOHN GAGE JUST HAPPENED BY THAT FEBRUARY AFTERNOON IN 1995 with a few innocent requests. He poked his head in the office door and asked for some computer cable and things. James Gosling rummaged around to get Gage what he wanted, but he then became curious: why did Gage need all this gear? Gage explained that he was giving a speech at a conference the next morning, and he wanted to finish with a flourish of high-tech flash. He planned to show off the software project Gosling had been working on for years at Sun Microsystems, a project that had never been shown in public. The software was in pretty solid shape, Gosling thought, but he was aghast at the prospect of Gage wrestling with it on his own in front of 500 people – a show that would depend on smoothly working Internet connections among other vagaries of technology.

"This had disaster written all over it, so I just got in the car with John," Gosling recalled. He went off to the conference in Monterey, over the hills and south of Sun's Silicon Valley offices, with no change of clothes, no toothbrush. He called his wife to tell her he wasn't coming home for a couple of days.

John Gage has the resume of a 1960s renaissance man – an All-American swimmer and lapsed math major at Berkeley who got enamored of the Free Speech Movement, made the Nixon White House "enemies" list, became deputy press secretary for George McGovern's quixotic presidential bid, attended the Harvard Business School and Harvard's Kennedy School of Government, and eventually returned to Berkeley. There he met the Berkeley Unix culture, which felt to him like a liberating force in computing, a technology movement that echoed his liberal political sentiments. Gage joined the corporate spinoff from the Berkeley Unix culture, Sun Microsystems, in 1982, two months after it was founded. He was employee number 21.

In recent years at Sun, Gage has been the director of the company's science office. His job involves traveling the world, looking for new developments and emerging ideas in technology, and he is also an emissary to the outside world, an evangelist for Sun's technology. He is an elite breed of marketer.

Gosling "just got in the car with John" that day, regarding the impromptu trip as another invigorating, half-out-of-control adventure with Gage. And, as Gosling suspected, the Internet connections at the Monterey Convention Center were untested and buggy. He spent all night and until 5 A.M. getting things in working order, much of that time on the phone with the engineers at MCI delivering a tutorial on how to set up the software on their routers. The next day Gage delivered his speech about the growing role of technology in education, but at the end Gosling joined Gage on the stage. With Gage supplying commentary, Gosling tapped into a Web page that held the image of a giant molecule, colored red, with three-dimensional shading. Nothing new there, but then Gosling began manipulating the balls – grabbing them, spinning them – bouncing them, all with a few clicks and strokes of his computer mouse.

Richard Saul Wurman, who runs the annual TED (Technology, Entertainment, and Design) conference, recalled that there was not much of a reaction to the presentation at the time. "Later, of course," he added, "everyone who was there said, 'Oh my God, I saw it there first.'" TED is an elite Hollywood-meets-Silicon Valley gathering. The TED crowd includes filmmakers, artists, authors, publishers, and musicians, as well as a smattering of techies. Oliver Stone or Quincy Jones are as likely to be invited to speak as John Gage. Yet for those who truly understood the Web in early 1995 – and there were not many at the TED conference – the software on display was startling. When he spoke to a few of them later, Gosling recalled, "They went, 'Holy shit.' It changed their world view, or at least of the Web as a medium."

Until then, the Web had been mostly a vast library of text pages and pictures in cyberspace. But the software Gosling put through its paces that day enabled *programs* delivered over the Web to run on anyone's desktop computer, anywhere. It held the potential to transform the Web from a medium of static Web pages into interactive programs. "It's as if you take a book, open it up and the page talks to you and you can move the characters around, and make them do whatever you want them to do," Gosling explained. A lively metaphor, but one that mainly conveys what it means to make Web pages – the most visible part of the Internet medium – programmable. More significant economically is being able to program the Internet, a low-cost global communications medium, to bring greater efficiency, speed, and diversity to all the transactions among companies and consumers in the modern economy. It is the broader meaning of so-called e-commerce, and it is just beginning. And it is Gosling's software, called Java, that has been the main tool for making the Web *programmable*, opening up a range of new possibilities and uses for the Internet, similar to the way that the stored-program concept expanded the horizons of computing itself by making computers general-purpose machines that could be programmed for all manner of uses.

Today, Java has become the Internet programming language of choice, the preferred teaching language in many universities, and the language of fluency for a generation of young programmers. In some ways, Java is the FORTRAN of the Internet age – a programming language, developed and championed by a single company, that is seen by programmers as a skill of the future. A team at Sun worked on Java, but its creator is James Gosling.

There are those who question his achievement. The technical ideas found in Java are all features that have been discussed in the programming community for a decade or two. So, they say, Java is not really new. But while the technical elements found in Java may have been lying around the halls of computer science for years, Gosling put them together in a working programming language for the Internet. All innovation is incremental in that it builds on top of previous knowledge. Yet a creative insight is required to assemble the building blocks of previous knowledge in new ways. Java contains that spark of fresh, organizing insight. "James was uniquely qualified to face the problem, and he had the deep understanding and taste to do a clean, elegant job," observed Raj Reddy, a computer science professor at Carnegie Mellon University, where Gosling earned his Ph.D.

Java is full of computer science ideas, but it is also a language designed around shrewd practical choices. In its blend of the intellectual and the practical, the product reflects its creator. Gosling is a fellow in Sun's research arm, and a corporate vice president, a title he wears with a characteristic informality. He stands a burly six feet, three inches, bearded and balding, with chestnut hair that falls to his shoulders. His appearance suggests an aging Viking warrior, though one dressed in a t-shirt and jeans, his standard attire around the offices of Sun's research lab in Mountain View, California

Gosling grew up in Calgary, Alberta, the eldest of three children. His mother, Joyce, was a high-school home economics teacher, while his father David had a variety of jobs. He sold oil field equipment and real estate, worked on construction crews, but mostly he handled the logistics for oil and mineral exploration ventures in remote locations far to the north. His father, Gosling recalls, was often "a disembodied voice at the end of a radio telephone" calling from places like Baffin Island or Tuktoyaktuk. To Gosling, his father's career was an entrepreneurial reality story, small-scale and struggling. "Dad was always looking for a deal, and he never really had a big winner," he said. "The primary breadwinner was always my Mom."

Watching his father's travails made a lasting impression on Gosling, leaving him with what he calls "sort of an entrepreneurial allergy." He explained, "I love to make things and see them work. But as far as the entrepreneurial things goes, it doesn't have the pull on me that it does on most people in Silicon Valley."

His fondness for making things, tinkering, and repairing surfaced early. He spent a lot of time, especially during summer, with relatives on their nearby farms. His grandfather Arthur had "this big boneyard of old farm equipment," he recalled, and starting at about age six Gosling found he was fascinated by the challenge of trying to stir some decrepit baler, thresher, or tractor to life. By the time he was 10 or 11, he had developed a knack for repairing things. At 12, Gosling made his first primitive computing device, with the unknowing assistance of the Alberta provincial telephone company. He scrounged through the trash canisters behind the phone company offices – a preteen adventure in dumpster diving – and found some phone relays and switches. Gosling used the parts, added a few embellishments, and built a contraption of flashing lights, switches, and plywood that played tic-tac-toe – the game that has introduced many programmers over the years to the basic challenge of building logic trees of if-then, and-or calculations. Gosling won a local science fair with his tic-tac-toe machine.

The family home was about a mile from the University of Calgary, where Gosling first became immersed in computer programming. His father took him over, and as they walked through one building Gosling noticed that some rooms were locked off. To get in, people had to punch the numbers of a security code. His curiosity whetted, Gosling went back later, hung around and watched people punch in their codes and memorized the sequences. "I got pretty good at that," he recalled. "And people figured that if you got in, you belonged there."

The building housed the university computer center. As a teenager, he became the computing equivalent of a gym rat – an eager, curious kid, always hanging around. And this kid, it turned out, had the mental muscles found in gifted programmers. Gosling was good at it, and he loved it, curiously fascinated by software as he had been by his grandfather's farm equipment, only more so. "In a lot of ways, software is the most complicated gizmo there is," he explained.

At age 15, Gosling was given a part-time job in the physics department, writing code to help analyze data from the Canadian ISIS-II satellite. The satellite was gathering information on the impact the aurora borealis had on international radio communications. Gosling's contribution to this showcase research project of the Canadian space program was to labor in the engineering trenches, writing, he recalled, "buckets and buckets" of line-at-time assembly code for a Digital Equipment PDP-8 minicomputer.

In high school, Gosling would often skip classes and sneak off to work in the university computer lab, a practice that brought reprimands from the school authorities but was quietly encouraged by Gosling's math and science teachers. "I was a good student of the annoying flavor," he observed. Gosling displayed the same tendencies at the University of Calgary, where he majored in computer science and apparently spent too little time in the rest of the university. Gosling did graduate in 1977, but it took the personal intervention of Anton Colijn, head of the computer science department, to ensure that Gosling was not denied a bachelor's degree merely because he sidestepped a curriculum requirement or two.

The usual rules were also bent somewhat to admit Gosling to the graduate program at Carnegie Mellon University. Gosling had applied to the leading American graduate schools in computer science: Stanford, MIT and Berkeley, in addition to Carnegie Mellon. Only the Pittsburgh school admitted him, and the nod from Carnegie Mellon was a bit arbitrary, as Gosling was told by a professor at the customary cocktail party following his defense of his thesis, "The Algebraic Manipulation of Constraints."

In selecting the 15 students for the graduate program each year from a couple thousand applicants, Carnegie Mellon, like the other schools, relied on recommendations from well-known computer scientists. Coming from the University of Calgary – a place not on the computer science map – Gosling was at a decided disadvantage. In the standard admissions procedure, Gosling did not make it. But each year, one student was chosen outside the usual process, mainly at the discretion of Joseph Traub, the head of the department. In his year, Gosling was told at the cocktail party, he was that student. "I didn't make it, at least not in the normal process," he said. "So in some sense, I won the lottery."

In 1979, Traub left Carnegie Mellon to head the computer science department at Columbia University, so he was not around when Gosling got his doctorate in 1983. Eighteen years later, in his Manhattan apartment that overlooks the Hudson River, Traub said, smiling, that Gosling's lottery analogy was a bit of an exaggeration. Traub did not recall precisely, but he said Gosling had probably done some rigorous research as an undergraduate that impressed him. "But sure," Traub added, "we took chances on people."

At Carnegie Mellon, Gosling was remembered by faculty and classmates as a mature young man who soon developed a reputation as someone with a special aptitude for software. "He was known to be the best programmer in my class, and probably in the whole school," observed Satish Gupta, a classmate. Even in his first year, Gosling would take on very ambitious projects, writing programs with tens of thousands of lines of code, far larger and more complicated than other first-year graduate students, and Gosling would do it over a weekend, recalled Gupta.

As a star programmer, Gosling shouldered much of the work in helping the school move to Unix, the powerful yet flexible operating system developed at Bell Labs. Aided greatly by the Bell Labs practice of licensing Unix to universities inexpensively and with few restraints on its use and modification, Unix became the software environment of choice at elite computer science schools like Berkeley and Carnegie Mellon. It gave young researchers room to roam and innovate. It also meant that Gosling, Carnegie Mellon's Unix ace, was regularly exchanging e-mail with a Berkeley Unix wizard named Bill Joy, who became a founder of Sun Microsystems.

When the Digital Equipment VAX machine, a powerful new generation of minicomputer, came out in the late 1970s, Carnegie Mellon had decided to run Unix on it instead of Digital's new operating system, called VMS. At the time, many of the school's programs had been running on an older vintage minicomputer, and Raj Reddy asked Gosling if he could get the old programs running on the new machine. The straightforward approach would have been for Gosling to write an interpreter – software that would take programs from the old machine, nipping and tucking line-by-line so that they would run on the new machine. Instead, Gosling took a more ambitious approach by building a machine code translator, a higher-level, more automated program. It handled the job at hand, enabling software written for the old minicomputr to run on the new VAX, but it was a more general solution that could be used to allow programs to run on many different kinds of machines.

Gosling says his work at Carnegie Mellon was the technical foundation for what became the "Java virtual machine." The virtual machine, as its name suggests, is a software component that acts much like a tiny computer. A programmer can write an application to run on the virtual machine, and the virtual machine can then translate the code to run on different kinds of machines, or on different operating systems.

Gosling began thinking about that kind of machine-independent software back in Pittsburgh in the late 1970s. "Obviously, that set of ideas stayed with James," said Reddy. "They helped him to create the small, compact virtual machine that could be carried with the code across the Internet. That was the problem he faced with Java – how to make code that is universal, that travels over the Internet and that can run on any machine. James had been working on these ideas for years."

Gosling took a year off from Carnegie Mellon, mainly, he said, to get some distance from a relationship "gone sour." He moved to San Francisco and worked as a consultant to Silicon Valley companies, mostly on their e-mail systems, before he returned to Pittsburgh, got his Ph.D., and went looking for a full-time job.

At the time, Sun Microsystems was just getting off the ground. Two of Gosling's friends from the university Unix culture, Bill Joy of Berkeley and Andy Bechtolsheim of Stanford, were founders. Both tried to persuade Gosling to join them, but he declined, owing in part to his "entrepreneurial allergy." Yet he was also being wooed by IBM to write software for a planned new line of work stations used mainly by engineers and scientists for computer-aided design and research – the same market that fledgling Sun was hoping to exploit. "IBM research group had some really cool stuff. They had built some prototypes that completely demolished anything Andy and his people were trying to do," Gosling recalled. "I was convinced Sun was going to be dead in months."

Working for IBM gave Gosling a foot soldier's view of how good technology can be stifled by bureaucracy and politics. The work station that resulted as a byproduct of the corporate skirmishes, Gosling said, was "big and it made a great door stop. I just couldn't believe they were being so stupid." After a year at IBM, Gosling fled for Silicon Valley and Sun. Shortly after he arrived in 1984, Gosling started an adventurous project known as NeWS, for Network Extensible Windowing System. His idea, radical at the time, was that distributed software could enable any computer screen on a network to display a program running on any other computer on the network. Other software experts were skeptical initially, but NeWs worked largely due to Gosling's uncanny knack for seeing ways to make code extremely compact and fast – essential for converting the idea of mobile code into reality. Though a technical success, NeWS never caught on as a commercial product. Still, it demonstrated the potential of exchanging code across a network, a potential that would Gosling realize with Java.

Java had its genesis in 1990. A young programmer, Patrick Naughton, announced that he planned to quit Sun to join NeXT Computer Inc., the maker of work stations started by Steve Jobs after he was pushed out of Apple. NeXT was the hotter technology company, Naughton told anyone who would listen. Scott McNealy, the Sun chairman, was listening and he asked the disgruntled young programmer to write him a memo detailing where he thought Sun was going astray. Naughton's "primal scream" e-mail, as Gosling called it, was circulated widely around Sun. Though certainly not a blueprint for action, it served as the catalyst for a lot of discussion within Sun. It tapped a nerve within the company's technical ranks, where there was a sense that Sun was stagnating, and needed to think big once again and stretch itself to create a great product for new computing markets. McNealy authorized the creation of a small team to chase that vision, including Gosling, Naughton (who was persuaded to stay with Sun), and a few others.

The idea was to pursue a long-range opportunity for Sun, looking out over the next 10 years. The broad trends in technology seemed reasonably clear. The steady march of microchip advances, according to Moore's Law, would mean that digital circuitry, computing power, and some element of intelligence could be placed in all kinds of devices before long. This ubiquitous computing would make sense only if the embedded computers could communicate over networks. The Sun group decided that the first market of sizable opportunity would be consumer electronics, where digital technology was already making inroads. They spent months doing their homework, and what they found was encouraging. The major companies – Sony, Sharp, Matsushita, Mitsubishi Electric, and others – all recognized that the worlds of consumer electronics and computing were converging. These companies were also very leery of Microsoft and Intel, who controlled the software and hardware standards in the personal-computer business, the closest cousin in computing to their own consumer electronics markets. They did not want to become technologically hostage to Microsoft and Intel, as the PC industry seemed to them.

For the Sun team, the next step was to try to develop some prototype technology and a business plan. Almost from the outset, the group moved off the Sun corporate campus, first into some unmarked rooms in the offices of a venture capital firm in Palo Alto and, in the spring of 1991, into a larger space, a floor above a branch of the Bank of America in Menlo Park. They worked in secret, even requiring other Sun employees to sign internal non-disclosure agreements to visit. The group's manager, Michael Sheridan, who would later leave Sun, recalled, "Looking back it sounds silly, but we were sure the Sun corporate 'antibodies' would kill us or get us de-funded if others knew what was happening." Their early product research included long stints playing Nintendo games. Before long, though, they settled on a goal: to build a kind of handheld butler for consumer electronics. With it, a person would be able control his television, video recorder, stereo, or anything else linked to its wireless network. It would have a touch-screen with icons for the devices and information sources, like *TV Guide*. To record a television program on the VCR, you would just drag the *TV Guide* listing onto the icon for your TV set, and drop it. And it would communicate; if your mother needed help, for example, she could e-mail the controls from her VCR to you. There would be no arcane commands, no keyboard; just turn it on and use your finger. Gosling's job was to develop the underlying software tools to make it work.

Back then, the project was code named Green. Its first business plan – "Behind the Green Door: Deep Thoughts on Business Opportunities in Consumer Electronics" – was remarkably prescient, at least about the requirements for the software that would become Java. (The title? Their office apparently had a green door, they were working in secret behind it, and the "symmetry" with the well-known 1970s pornographic film of the same name "appealed to our twisted senses of humor," Gosling recalled.) Marked "Sun Confidential and Proprietary," dated Aug. 23, 1991, the 44-page document described the technology under development, its importance, and the potential market for it. The report defined consumer electronics broadly to include "things like VCRs, telephones, games, automobiles, dishwashers, thermostats," and Green would supply the technology to "allow these devices to interoperate transparently." Green, the report added, was "primarily a software technology development group."

The report said that Sun should move swiftly to become the software standard in consumer electronics, just as Microsoft did in the personal-computer industry. "It is a chance to 'end run' Microsoft's dominance by going after the volume inherent in CE (consumer electronics) while standards and interfaces are still fluid and undefined," it said.

In the plan, there is a description of GreenTalk, a "robust programming" language. In two pages of small print, the report discusses GreenTalk as a language ideal for handling distributed mobile code, and enumerates several of the technical characteristics found in Java today, including the virtual machine concept of being able to "dynamically adapt" to different machine environments. "The big stuff in Java all comes out of that business plan," Gosling said years later.

In the GreenTalk days, Gosling may have had the consumer electronics market in mind, but the problem he was addressing was remarkably similar to the one Java addresses on the Internet: how to fashion software that can transport programs over vast networks to many users, who run those programs on all kinds of computing devices. He was thinking of networked computing that could flourish on its own in a messy world of diversity and unsophisticated users, as opposed to private corporate networks in 1991, isolated and constantly coddled by experts. "Once you move into the realm of networked computing with real people there are a cascade of things that become consequences of that world," Gosling said. "The design points in Java all come from confronting that problem."

Gosling's ideas came gradually at first, and evolved over time. When he began in 1990, he had no intention of creating a new programming language. "It really did start out as a fix here and there to C++ to make things work better," he recalled. "I was just trying to build code to make the project work, the handheld prototype. But the further I got into it, the more it became clear that I had to go deeper."

One moment of clarity, he recalled, came in 1991 during a Doobie Brothers concert in Mountain View. His sister Barbara was visiting, and she somehow managed to get front-row seats. The speakers, Gosling remembered, were slightly behind him, so he could actually hear the unamplified voices of the singers. Sitting with his feet perched on the stage, Gosling gazed up at a network of lights, robotically controlled, that darted around to the changing rhythms of the music. This sophisticated network of lightbulbs seemed a real-world analogy to the Sun team's theoretical discussions. "It triggered a lot of thinking" on the software design necessary to handle issues like security in a networked environment, Gosling said. "It's not something you can add on. It's something that goes down to the ground."

The "Behind the Green Door" plan was a stake in the ground – a statement of direction, and not a finished product by any means. Gosling worked throughout 1992 on his language, which he soon called Oak, a name inspired by the tree outside his office window. The handheld prototype, the Star7, was finished and shown to McNealy in September 1992. The touch-screen device was impressive, with its snappy icons, sound effects, and even a virtual guide dubbed Duke, a cartoon imp with a big red nose. McNealy was ecstatic and Sun set up a wholly owned subsidary, FirstPerson Inc., to market it. The consumer electronics companies were impressed, but they decided that Sun had put too much costly technology into its design to be a mass-market product. They said no. That rejection was soon forgotten, however, as another opportunity beckoned.

By the early 1993, the media, technology, and entertainment industries were in the thrall of the notion of the "information superhighway," which would be television, digitally reconstituted – 500 channels, movies on-demand, shopping services, newspapers, magazines, and books – all delivered to the screen in the den. In the spring, Time Warner set off a corporate stampede when it said it was seeking partners for an interactive television trial in Orlando. Technology companies fell over each other, hoping to supply the set-top boxes, while other cable companies hurriedly declared plans of their own.

The FirstPerson group, including Gosling, spent the next several months trying to convince Time Warner that Sun would be the best supplier of set-top boxes, with Oak as the ideal software medium for handling all that information, entertainment, and electronic commerce. The contract eventually went to Silicon Graphics, and the high-profile Orlando trial and other experiments in interactive television proved to be costly failures.

During the months spent pursuing the Orlando bid, Gosling was struck by the conceptual chasm between the Sun team and the cable company executives. All along, the information superhighway Gosling and his peers had in mind was "an Internet for the rest of us," instead of something suited for computer sophisticates as the Internet was in the early 1990s. In the Sun vision, the network would be open to all kinds of services, and control would mainly rest with the consumers. "We were talking about a fairly egalitarian, peer-to-peer kind of thing," Gosling recalled. "Their technical guys loved us. But the people at the real top of the cable companies hated it because their business was to be a gatekeeper of eyeballs. These guys had a serious addiction to control and we had been preaching anti-control."

The FirstPerson unit was folded in the spring of 1994, and Gosling's work was in danger of going down with it. In fact, Gosling was told later, there was a plan to cancel the Oak project and assign the programmers elsewhere. The plan was never carried out, thanks largely to Bill Joy. A Sun founder, Joy is the company's technology guru and a member of the company's executive committee. He is a brilliant computer scientist, but one with a decidedly catholic intellect. A conversation with Joy is apt to jump and loop over subjects including opera, literature, architecture, and the fine points of various Manhattan hotels, before coming back to some programming issue. In 1991, Joy moved from Silicon Valley to Aspen and, in his phrase, "left the urgent behind to get to the important." He set up a small Sun research-and-development lab in Aspen, a town whose amenities, he explained, include excellent skiing, hiking, and a first-rate local book store.

Since the beginning, Joy had been what one colleague termed a "rogue contributor" to the Oak project, focusing on it intensely for a couple of weeks and then disappearing for months. When it looked as if it might be canceled, Joy intervened. "Bill did a lot of serious screaming and yelling to resurrect it," Gosling said.

By 1994, the Internet wave was beginning to build. In June 1993, two students at the National Center for Supercomputing Applications at the University of Illinois, Marc Andreessen and Eric Bina, released the first version of the Mosaic browser. To commercialize the browser, Andreessen cofounded Netscape

Communications Corporation (called Mosaic initially) in April 1994, and Bina joined him. It didn't happen overnight, but easy-to-use software for browsing and viewing pages on-line would help transform Tim Berners-Lee's Web from a neat tool for researchers into a mass medium. The Web provided the technology for presenting, identifying, and linking on-line information, while the browser added an easy-to-use viewer, suddenly making the Web accessible to anyone with a personal computer. And Gosling's software could be adapted to become a powerful Internet programming language. At a corporate retreat in Lake Tahoe, Joy pressed the point. "I told them roughly that 'the game's afoot,' and that the game is the Internet," Joy recalled. Gosling and the others agreed. "We didn't think the Internet would ever become consumer oriented," he said. "But because of the Web and the browser it was starting to happen. So we just picked up our toys from one sandbox and moved to the next."

Joy had long been pushing the view that Gosling's language alone was a significant development in computing, even if it wasn't directly tied to some Sun hardware product that would generate sales and profits according to the company's tried-and-true business formula. Since the late 1970s, when he began modifying the Berkeley version of Unix for the early Internet, Joy had been thinking about the possibility of a programming language that would ensure greater safety and reliability in the emerging environment of diverse networks of computers. To do that would require deftly combining a series of advances in computer science over the years into a new language. "I didn't know how to do it," Joy said. "But I knew it when I saw it," and Java was it.

In Java, Joy recognized Gosling's special talent as a programmer. "James," he said, "is great at writing the first version of a completely unexplored space, mapping it partially but wonderfully. It is a rare skill, to be able to do this as a solo programmer. To make the whole thing fly for production use, others have to follow behind. But pioneers like James are rare. His is a rare form of imagination, to be sure."

The genius of Java, observed Eric Schmidt, the former chief technology officer of Sun, is that "James borrowed and invented in exactly the right ways."

To appreciate Gosling's software design choices – his borrowing and his innovation – it helps to explain a bit about the computing context from which Java arose. That context was defined both by what Java was moving toward, and what it was reacting against. Gosling's efforts were aimed at a networked world, where pieces of software mix, meet, and communicate with other pieces of software in increasingly complex patterns of interaction. This is in striking contrast to the non-networked setting, where programs live as if they were part of an isolated tribe in a remote land. But the safety of tribal isolation is lost once software is exposed to the wider networked world. The quirks or bugs in a program, harmless in a more isolated environment, can have effects that are unforeseen, damaging, and far-reaching in the networked world. The potential for the spread of troubles over the Internet is vividly illustrated by consciously malicious programs, viruses.

Java was also reacting against the dominant languages used for industrial-strength programming, C and its closely related successor C++. Gosling and his team used C++ as a guide as to what should be avoided in some respects. As a language, C++ gives programmers much freedom to manipulate data and computing instructions down to the bit level. It is a powerful tool, but a complicated tool, and often a one dangerous in the hands of all but very skilled programmers. "C++ is an M-16," Arthur Van Hoff said, referring to the military automatic rifle. A Dutch programming ace who joined Gosling in 1992, Van Hoff noted that from the outset the team had focused on building a "language with no sharp edges, so people couldn't hurt themselves. And remember, ninety percent of all programmers are pretty average programmers. So, especially in a networked environment, complexity is a killer."

Gosling offered a historical perspective. In 1972, when C was created at Bell Labs, it was developed to give programmers fine-grained control of the machine while working with the minuscule memory capacity of computers in those days. "C was a brilliant design for a language in its day," Gosling said. "But the things people care about who use code have changed. Computers weren't networked then. They were isolated behind locked doors. No one cared about code running on all different kinds of computers. You just hoped and prayed that your code would run on that big, slow machine in front of you."

Java, by contrast, is a language for the Internet age. It is designed to simplify the programmer's job in a complex, networked world. Java does this first by including tools to make the programmer's life easier and more efficient. One such tool is automatic memory management, or "garbage collection" in the jargon of programming. The garbage collector, in this sense, is a software snippet that searches a computer's memory for dormant data and program segments, shoves them out, and then reclaims that memory space for more pressing work.

C and C++ do not include garbage collection as a standard feature, which means the programmer must handle memory management by hand, keeping lists and tracking down dawdling data. It is a time-consuming distraction from more useful work in the Internet environment, Gosling decided. And house-keeping lapses in memory handling often cause software to misbehave. Besides, computing power had increased so much that running a garbage collection routine would no longer really slow a machine's performance. So what might be seen as an impractical frill earlier had scant, if any, penalty for most users when Gosling was designing Java in the 1990s. "Twenty years of Moore's Law made a huge difference," he observed.

Java also tries to simplify Internet programming by limiting the programmer's freedom to make mistakes. There are many such restrictions in Java, and they are often deeply technical matters, but a couple of general examples illustrate this side of Java's culture. For example, Java is quite strict in its treatment of "exceptions," or aberrant behavior by a program. When software is being compiled – translated into code the machine can read – a programmer may be alerted by an "exception report," or warning before the program actually runs on the computer.

In other languages, these warnings are typically advisories, and can be ignored – indeed, exceptions need not cause problems. Yet, in a networked world, an oddly behaving program is more likely to cause trouble. Java insists that the programmer take some action in response to these exception warnings. It is a language that forces programmers to think about places where things can go wrong. It does so in an effort to make the resulting software more reliable in the complex, diverse software environment of the Internet.

A defining characteristic of Java is that it is what computer scientists call a "strongly typed" language. Simply put, this means that the kinds of data, and the operations that can be performed on each kind, are tightly classified. In Java, the software modules are strictly defined, as are their interfaces, or hooks, that allow them to interact with certain blocks of software and not others.

The ideal in the Internet world, says Joy, is to try to ship software that is as reliable as a piece of machinery. Java can then be seen as a tool kit of screws, nuts, bolts, levers, and gears. Its strong typing is a system to ensure that the programmer does not try to do anything chancy with the tools. Java, to be sure, is a "systems" language, meaning that, like C, C++, and Lisp, it enables the professional programmer to reach down deeply into the machine to manipulate the bits, and thus alter the basic operations of the computer. But Java also enforces structured guidelines for doing so, without wiggle room or backdoors. It takes away hazards, but it also takes away freedom.

Not everyone is comfortable with restrictions Java places on programmers. Programmers who resent its strictures regard Java as a police state, and Gosling as a software fascist. Gosling replies that the disciplines of Java are assets in the Internet environment, freeing programmers from a lot of detail work and ensuring that software is more reliable. Sure, he says, Java is a tradeoff, but one in which the benefits far outweigh the costs.

"Java is a language where the rules are the rules," Gosling said. "Once you adapt, it is really an incredibly liberating thing." He compares the criticism of Java with the outcry heard from early fighter pilots, irritated when aircraft manufacturers sealed off cockpits. In the old days of propeller-powered planes, pilots stuck their heads out to navigate and sniff the air, sensing the winds and the weather. "But when you are in a plane that is flying at Mach 3, if you open up the cockpit to look out you'll get your head ripped off," he said, warming to his programming point. "To free yourself up for the next level of scale, you have to give up things that used to feel like freedom."

By 1994, Gosling had the major concepts in place, but much work remained to craft Java into a working language for Internet programming. While Java eventually proved to be the right technology with the right timing to ride the Internet wave, none of that was evident back then. A successful technology tends to look inevitable only in hindsight. In particular, the way Java could meet the developing need for an Internet programming language was not widely apparent, because the rapid evolution of the Web into a mainstream medium of commerce and communication had scarcely begun.

Guy Steele joined Sun in the fall of 1994 from the Thinking Machines Corporation, an innovative but commercially faltering supercomputer maker. A well-known computer scientist, educated at Harvard and MIT, Steele is an expert in computer languages and artificial intelligence. After joining Sun, Steele made a survey of the research projects at the company. There were a couple of "C replacement languages" under development, including Oak, as Java was then called. Steele viewed Oak as a tightly written language that had stripped out elements that he termed "the clutter of C ++," while adding garbage collection and a few other things. "But I wasn't that impressed with Oak," he recalled. "I didn't see the applications until later. I had a blind spot."

Over the next year, a series of defining decisions had to be made about Java – what features would go in, what would be jettisoned, precisely how to build each feature in, and how the ingredients would fit together. Moreover, the code had to be honed to improve its performance. The key decisions were made during a series of meetings in SiliconValley and Aspen. They were typically attended by about a half-dozen people. The matters may have been technical, but the debates were often passionate, even bitter at times, especially the shouting matches between James Gosling and Bill Joy. As chief technology officer, Eric Schmidt attended a couple of the Aspen meetings. "They were incredibly rough, sometimes very unpleasant," he recalled. "But while it may have seemed very personal, it is the way strong-willed, brilliant people work these things out."

The crux of the Gosling–Joy debates centered on the tradeoffs between simplicity of the language versus the power of the language. The Joy stance was that the group should try to make Java as powerful as possible by putting in a host of cutting-edge language features. Decades typically separate one major computing language from the next, Joy explains, noting that C came out of Bell Labs in the early 1970s, but that its parentage went back to BCPL in the 1960s. Once a major language is in place, changes tend to be glacial.

When he looked at Java, Joy saw an opportunity to reverse what he saw as the "trifurcation" of computer languages into numeric, systems, and symbolic languages. To explain, he pointed to the example of how people will use increasingly powerful handheld computers: for stock prices or scientific data, a numeric language like FORTRAN is best-suited to the task; for its display and network links, the device could use a systems language like C; to give it some intuitive intelligence for suggesting scheduling or shopping alternatives, the device could use the skills of a symbolic language like Lisp. Joy hoped to put the strengths of them all in Java. "I viewed Java as a once in 30 or 40 years chance," he recalled. "So I tried to get everything in."

Gosling was the advocate of simplicity. His thinking boiled down to, "When in doubt, leave it out." Reducing the complexity, he figured, would make Java a more coherent, small, and friendly language, and thus more understandable and inviting to programmers who would have to be won over. The simpler the language, Gosling recognized, the faster it could be shipped, capitalizing on the window of opportunity suddenly being created by the Web and Netscape's browser. Joy's role, according to Gosling, was as a useful sounding board, pushing hard on several points, even getting his way a few times. "But James was the ultimate arbiter," said Guy Steele, a regular at the meetings. "It was his baby. And we all felt that it was important that the language be consistent technically – that it be one person's vision." Java would not be a programming language by committee. "Looking back on it," Steele added, "James made very much the right decisions. He made a tasteful selection of what to leave in and what to leave out. And what he wanted turned out to be important."

The selection shows the practical streak in a computer scientist who first displayed his aptitude for engineering in Calgary, fixing farm machinery. When asked what he thinks makes a programming language a winner, not just Java but the others over the years, he acknowledged that luck plays a role. But mainly, Gosling noted, each of the major languages have been the most useful tool of its time. "In the language design community, there is a lot of academic debate and thrashing, all of which has the flavor of how many angels can dance on the head of a pin," he observed. "But in my naïve view of the world, it actually has to solve real people's problems, and it must do so without changing everything."

One of Gosling's shrewdly pragmatic moves was to make Java look instantly familiar to millions of programmers. The source code – the top layer of a language that can be read by humans – is rendered in the vernacular of C++. So to legions of developers worldwide, Java appeared to be a language they could pick up easily by adding a few wrinkles to their hard-won knowledge of C++. That appearance is somewhat misleading, but it surely hastened the pace at which Java was adopted. Also like C++, Java is made up of small software modules, or building blocks, known as objects. Yet Java mostly builds on ideas from other languages like garbage collection, byte-coded interpreters (the enabling technology for the virtual machine), and strict definitions for data types and rules of behavior.

"To the developer, it all kind of looks like C++," Gosling explained. "But underneath the sheets, Java owes a lot to languages like Lisp, Smalltalk and Pascal. What I did was kind of smoosh these environments together."

While the language was being crafted into shape, so was yet another business plan. Though rescued from extinction in the spring of 1994, the Java project still lacked direction. Gosling, Joy, and others agreed that the Internet was now the market opportunity, but how to exploit it? At a strategy meeting in September, Schmidt the chief technology officer, and Joy met with Gosling and other members of the Java team. The subject was goals and objectives. When Schmidt asked, Gosling replied that what he wanted was "everybody using the language."

The business plan Schmidt drew up immediately after the strategy meeting embraced Gosling's goal and put a number on it – a target of 100 million users of Java software within five years. The 100-million figure was used as the benchmark for "everybody" because, Schmidt recalls, that was the number of desktops that ran Microsoft's Windows operating system at the time. In fact, Java would reach the 100 million threshold not in five years, but in two. Sun accomplished that with perhaps the most aggressive and effective marketing campaign a new programming language has ever enjoyed – a campaign not merely intended to win the hearts and minds of programmers, but also top management, the press, and even the public. The effect was to create what one industry executive called an "an aura of inevitability" surrounding Java, so that the whole computer industry – even Microsoft – eventually agreed to distribute Java.

The biggest single step came when Netscape endorsed Java, and agreed to embed it into its browser. At the startup, the Netscape founders understood that the browser was only a first step, making picture-and-text Web pages easy to retrieve and view. But if the Web was really going to become an everyday medium of commerce and entertainment, it needed the real power of computing. "We needed a way to make the Web programmable," recalled Marc Andreessen, cofounder of Netscape.

Indeed, Andreessen and the other young Internet programmers at Netscape had been mulling the issue for months: What would be the best programming vehicle for handling mobile code sent from big hub computers, or servers, over the Internet to millions of desktop machines? They looked at adapting Lisp and Scheme, a Lisp variant, for the task, but those languages came tainted with reputations as specialist, niche languages, difficult to read and understand. Andreessen and his team were then shown Java, and they liked what they saw. It seemed to have the necessary technical ingredients, and it looked familiar on top. "Using the C++ syntax was brilliant," Andreessen said. "It meant you had a ghost of a chance of getting it adopted by the programming community."

Schmidt and Joy negotiated the deal with Netscape, which signed a letter of intent on May 23, 1995. By then, the ball was rolling. Gosling's Oak had acquired a name with more marketing pizzazz, Java. Gosling recalled the name-winnowing process. He and the rest of the project team were placed in a room with a white board and meeting facilitator, whose principal guiding question was, How does this technology make you feel? Three hours later, they emerged with about a dozen candidate names. At the top of the list was Silk, which

Gosling hated. Ranked third was Lyric – Gosling's personal favorite, he says, because programming languages, like song lyrics, are "the way you express yourself." Fourth was Java, since someone mentioned that it made him feel excited, as if he had overdosed on coffee. "That was the extent of the connection," Gosling said shrugging. The candidate names were sent to the Sun legal staff for trademark review.

Soon afterward, Kim Polese presented two candidate names, Silk and Java, at a corporate staff meeting presided over by Schmidt. Schmidt told Polese, the product manager for the new programming language, that it was her decision. She made her choice, and Java it was – a far cry from an oak tree contemplated outside Gosling's window.

The Internet business plan for Java was a departure for Sun. Joy and Schmidt were guiding the strategy. McNealy gave them a lot of leeway, and they kept him informed. He knew they signed the letter of intent with Netscape, and he approved. Yet what Joy and Schmidt had not spelled out to Sun's top management and its board was that their plan did not include making money from Java, at least not directly. The licensing terms were all but a giveaway, intended to accelerate the adoption rate of the software. And since programs written in Java were intended to run on any computer and operating system, it seemed to hold the danger of hurting Sun as much as Microsoft. For Sun, like Microsoft, made a lucrative living by "locking in" customers to its technology – Sun computers running its Solaris operating system.

The Java strategy was a bet that Sun would nonetheless benefit from establishing Java as an Internet standard. The bet would prove to be a good one, but it was a leap of faith more than a conventional business plan. When he discussed the Java plan informally with other Sun executives, Schmidt recalled that the reception ranged from skepticism to outright antagonism. Joy got a similarly chilly response. "The most difficult thing was to introduce a new order of things, classical Machiavelli, in an organization so focused on another model," he said. Yet in June 1995, when Joy and Schmidt presented their plan for Java to McNealy in detail, he liked it immediately. "Scott understood all the implications, and he really drove the strategy himself after that," Schmidt recalled. "Bill Joy and I delivered the idea, but Scott delivered the company."

Sun then embarked in earnest on the long march to make Java the industry standard for Internet programming by building a commercial ecosystem of more and more programmers trained in Java, writing more and more Java applications. By 2001, Java was licensed by more than 200 companies and used by a couple million software developers worldwide. Many people believe that Java will be the dominant systems programming language on the Internet for the next 20 or 30 years, though not everyone agrees, particularly Microsoft.

In January 2001, Microsoft settled a long-running suit with Sun, agreeing to pay Sun $20 million and to terminate its license to use Java. In the suit, Sun charged that Microsoft had violated the terms of its license by adding special extensions to Java for its Windows operating system. Sun claimed this was a blatant attempt by Microsoft to "pollute" Java and fragment the standard by undermining the essential principle that programs written in Java can work well with different operating systems. Microsoft countered that it had done nothing wrong, asserting that Sun brought the suit as a competitive tactic to "disadvantage" Microsoft. In any case, a couple days after the settlement, Microsoft announced a "Java user migration path" to Microsoft technology. And it is promoting its new programming language, C#, or "C-sharp," as an alternative to Java. Microsoft is positioning C# as a language that does all that Java does, with fewer restrictions imposed on programmers. It is textbook Microsoft – embracing a technology, redefining it on its own terms and then persuading millions of programmers to join its parade. "We think the time is right for another language," said Anders Hejlsberg, a senior software designer at Microsoft. "In the C family alone, we've had C, C++ and Java. What makes us think that Java will be the last one?"

Hejlsberg could be right, and Microsoft's C# promises to be a formidable contender. Still, Java seems well ahead today. Java has become the teaching language at more and more universities. It is the preferred language of the developers who are the carpenters and bricklayers of the Internet economy – the programmers building the transaction systems and industrial-strength Web sites for on-line commerce. When asked about his staff, Rich Buchanan offered the standard answer. "They're hard-core Java programmers," said Buchanan, engineering director for a firm in New York that builds and maintains large corporate Web sites. "Java is where software engineers want to be working these days. It makes them marketable. It's seen as a skill of the future."

The lessons of Java are several. A good technology, timing, luck, and marketing all played a role in its success. There were ups and downs, fits and starts. But Sun got Java because it gave a small team of its best engineers the freedom to go off exploring for something new and significant, and ultimately listened to James Gosling when he found it. He not only had something to say, but he knew how to say it, so his advocacy for his technology was seen as persistence instead of stubbornness or mere self-involvement. "You've got to be able to explain things to a wider audience in ways they relate to," he noted, "It can't be explained in ways that are so esoteric that no one gets it. That's the engineer's weakness – insular thinking."

Yet Gosling is also unabashedly "geeky," in the sense of being deeply fascinated by machines and gadgetry, and by the riddle of how things work and why. On a ride to lunch, Gosling clearly enjoys putting the new navigation system in his gray Volvo convertible through its paces. It not only tracks the car on a digital screen and offers guidance, but is also linked to a database that, among other things, sorts through nearby restaurants by type, from French to Thai. "Now, it's going to get angry," Gosling said, grinning, as he ignored the machine's suggestions and turned off in the direction of a favorite Indian restaurant.

The gifted programmers, Gosling says, have the geek gene. "There is an odd and obsessive side to it," he explained. "The people who are best at it have a temperament that makes them the kind of people who are intellectually drawn to something like it's magnetic, sucked into it, and they don't quite know why."

With Java flourishing, Gosling has gone back to exploring in the labs, working to create new software development tools that he calls "the saws and hammers of the craft." The tools Gosling is trying to fashion would be to help elite programmers grapple with the "complexity crisis" in software, as more and more programs must cooperate and behave over the Internet. His current work, he says, shares a common thread with Java, which for Gosling began as an effort to make the software tools needed to build a product. "My whole career has been that way," he said. "I start out to build something and I find it would be really helpful to have better tools. So I start making the tools. Soon, you find yourself going deeper and deeper, down into the muck with the tools, and you've forgotten about that thing you started out to build in the first place."

11

There Has to Be a Better Way: Apache and the Open Source Movement

THEY FELT ABANDONED IN LATE 1994 – a small, scattered collection of Internet software engineers from Britain, Nebraska, San Francisco, and elsewhere, all commiserating on-line. They were mostly people running Web sites in the days before the Internet was a household word, before Wall Street took notice, before the browser war. The Internet was still a clubby little realm whose practices were guided by the research culture of its origins in the late 1960s. In the tradition of academic research, Internet software was freely shared.

But things were beginning to change. The programming team from the National Center for Supercomputing Applications at the University of Illinois had just decamped to Silicon Valley to seek their fortune, joining a startup that would soon be called Netscape. The on-line commiserators all used the same data-serving software, developed at the Illinois supercomputing center, for delivering Web pages over the Internet to desktop computers. With the Illinois developers gone, the far-flung group of Web masters shared code improvements and bug fixes informally among themselves. "We traded software patches like baseball cards," recalled Brian Behlendorf, one of the original group.

Lacking a leader or an organized approach, however, the accumulation of quick fixes threatened to soon become a patchwork mess of a program. So eight of the software engineers got together and engineered a process. "We decided to take the code we had and start our own project," said Randy Terbush, another member of the original group. The so-called server program, they agreed, should be designed and built in well-defined software modules, so that individual programmers could easily work to burnish one building block of code without having to worry about the entire program. A governing process was set up that mirrored the design philosophy of simplicity — features could be added, but not unless the need was clear and the members agreed. They called their collective effort Apache — a name derived from a self-deprecating joke among the developers at first that theirs was "a patchy" server.

Since early 1995, the Apache server software has been rewritten time and again, and the program itself has grown to more than 400,000 lines of code. People have come and gone. Competing programs have been made and marketed by companies with skilled programmers and deep pockets. But the Apache software — distributed free, coded and debugged by a worldwide group of volunteers — is the clear leader in its niche. In the spring of 2001, Apache ran on more than 60 percent of the computers serving up Web pages over the Internet, while the rival offerings from Microsoft and iPlanet (an alliance of Sun Microsystems and Netscape, which was acquired by America Online in 1999) trailed far behind. Apache, its developers acknowledge, may not be the fastest Web server or offer the richest set of features, "but we have the momentum," said Brian Behlendorf.

Apache is one of the showcase success stories of "open source software," a term that represents both a philosophy and a software development model. In open-source projects such as Apache and Linux, software is distributed free and its "source code" — the coded instructions that a knowledgeable programmer can read and understand — are published openly so that other programmers can study, share, and modify the author's work. The open-source model represents a sharp break with the dominant approach of most software companies today, in which source code is regarded as the supplier's private property. In the industry, software is generally distributed not as source code but in binary form — the 1's and 0's executable by the machine, but not really comprehensible to humans, especially given that modern programs can total hundreds of thousands or millions of lines of instructions. Under strict licensing terms, software companies will sometimes share source code with their favored corporate customers, but such sharing is typically done so big customers can view and understand the code, not so they can modify the code, as in the open source model.

The open-source "movement," as it is often called, is a combination of idealism, software engineering, and business tactics, enabled by the rapid spread of the Internet in the 1990s. Its heritage of idealism owes much to Richard M. Stallman, a star programmer at MIT's Artificial Intelligence Lab during the 1970s, who later founded the Free Software Foundation. All software, according to Stallman, should be "free" in the academic sense of freely-published research, subjecting work to peer review and sharing one another's discoveries. To Stallman, the rise of "closed source" software in the 1970s, when software began to be widely sold separately from hardware as a proprietary product, was a mistake – a triumph of greed over morality.

The pragmatic, "software engineering" argument for open source is that the Internet, as a medium of low-cost, immediate global communication, has fundamentally altered the dynamics of software development. The Internet, this reasoning goes, allows programmers to share ideas, suggest improvements, and fix bugs in a far more productive way than could be done in the closed-shop setting inside a single company. The engineering champions of open source believe the world's best programmers will coalesce around the software challenges that interest them, each contributing, stimulated by his or her peers, and a "meritocracy of code" will result. In the Internet era, they assert, many eyes serve to accelerate the pace of software development, in sharp contrast to the dictum of the software-engineering guru Frederick Brooks, who observed that adding more programmers to a big project will only slow things down. In his essay "The Magic Cauldron," Eric S. Raymond, an open-source advocate, declared that "decentralized cooperative software development effectively overturns Brooks's Law, leading to unprecedented levels of reliability and quality on individual projects."

Open source is also a business strategy that seeks to change the terms of trade in the software industry, taking power and wealth away from the dominant suppliers, especially Microsoft. To pursue that goal, the open-source insurgents began with a makeover and a marketing campaign. In 1998, they adopted the term "open source" instead of "free" to describe their development model, because speaking of "free software" seemed to suggest they were anti-capitalist extremists. In the open-source vision, there will still be plenty of money to be made in software, but software will become a service business, instead of the traditional model of shipping proprietary programs as if they were manufactured goods. The service fees would go mainly for helping individual and corporate customers *use* software, instead of seeing the industry's wealth go into the pockets of those who *sell* software. Open-source enthusiasts insist that the shift in power from sellers to users is only fair. After all, they note, only 20 percent of professional programmers work for software vendors who *sell* software, while the other 80 percent work as in-house programmers in companies, or as consultants – all trying to help people *use* software.

The sensible software efficiency message of open source is appealing. Major companies – and IBM, in particular – have supported and invested in the two leading open source projects, Apache and Linux. Open source is doing well in certain cutting-edge markets, such as Internet servers, Hollywood special-effects and animation, and supercomputing. Europe sees open source as way to close its software gap with the United States. Computer science students in universities think open source is cool, since it seems to combine the fun of programming and the camaraderie of team play with a certain renegade satisfaction of siding with the outsider, the underdog. Microsoft is concerned, and in the spring of 2001 dispatched one of its top executives to speak at universities and industry gatherings, extolling the virtues of the "commercial software model." Just how mainstream open source will eventually become is uncertain, but it is beginning to build a diverse ecosystem of self-interest.

Brian Behlendorf epitomizes the pragmatic wing of the open-source movement. Indeed, he avoids using the term "movement" when discussing the open-source phenomenon, since it connotes a political movement instead of an economic model. "If open source is to succeed, it has to have a business justification," Behlendorf explained. "A lot of people are in the open-source community because they think it is the right thing to do. But charity only goes so far. You've got to make it sustainable." Behlendorf, whose hair drops to his shoulder blades, was padding around in his socks in the loft-like offices of Collab.Net, a San Francisco firm that helps companies design and run open source projects.

Behlendorf was born in 1973, and grew up in Southern California. His parents, Robert and Becky, met at IBM, where they were both systems engineers of the COBOL era. He grew up with personal computers around the house, and took programming classes starting in the fourth grade, when he learned Logo, an instructional language for children created by MIT's Seymour Papert. He later dabbled with a little BASIC programming, but by the sixth grade he "fell out of computing," he said. "It was the stigma, it seemed so arcane and uncool. And also, kids, I suppose, rebel against what their parents do." In high school he was a well-rounded student who ran cross-country, worked as a disk jockey at school events, and graduated near the top of his class. He got back into computing when he attended the University of California at Berkeley, because of his interest in music. Behlendorf got immersed in the electronic dance music played at all-night raves in San Francisco. It was a music scene populated mainly by university students, some of whom had access to Unix work stations and the Internet, and Behlendorf was soon running the Internet mailing list for "San Francisco raves" and keeping an on-line archive of music files.

In 1993, Behlendorf got a part-time job as a "$9-an-hour Unix sherpa" for a new magazine in San Francisco that wanted to set up an Internet account for e-mail and to put the text versions of its articles on the Web. In its early days, *Wired* magazine led the way in journalism in presenting the Internet as not merely a technology, but as a social and cultural force, and made that case with innovative, colorful graphics and passionate writing. In the first issue in 1993, founder Louis Rossetto declared "the Digital Revolution is whipping through our lives like a Bengali typhoon." Its message resonated with Behlendorf. Though an hourly contract worker, he recalled, "I felt like I was doing God's work in that first year and a half at *Wired*." For an older generation, the personal computer had been the bracing technology, with a whiff of social change clinging to it – a technological tool of individual empowerment. But Behlendorf's generation took PC's for granted, and for him computers only got really interesting when lashed to the Internet as tools of communication and community. The same could be said of most of the people who joined the Apache project. "Lot of us were born and bred of the Internet mindset," he said.

Behlendorf was 21 and still at *Wired* when the Apache project began. They started with the server code from the University of Illinois supercomputing center and set to work. "Being good engineers, we came up with a process with no single point of failure," Behlendorf said. "From day zero, it was a multilateral effort." To succeed, an open source project must have strong, coherent leadership. The Apache project has accomplished that with a group of "cardinals" instead of a single "pope," as Linux has in Linus Torvalds, its creator. Though tricky, the collective governance has worked so far because of the chemistry of the Apache group, their shared objectives, and a common philosophy. The goal was to craft a highly reliable Web server that could scale up quickly to handle high volumes of traffic. The Apache project was composed of Unix programmers who approached the design of the server with the Unix "tools" mentality, with its emphasis on simple software building blocks that fit together neatly. The simplicity was easier to maintain, according to Randy Terbush of the Apache group, because of the lack of commercial pressures. Features and fixes, Terbush noted, were "dictated out of necessity and not driven by attempts to win a marketing war."

The Apache group was not the source of the initial design of their server software. That was done mainly by Rob McCool at the Illinois supercomputing center, before he departed for Netscape. Indeed, innovative programs are typically conceived by a lone craftsman, or a few artists at most. But, as Eric Raymond, the house essayist of the open-source community, has artfully described it, the real strength of the open-source model comes when the original design is placed into an intellectual "bazaar" of programmers cooperating to improve, refine, and debug the code.

Since the code is in the public domain, there is no legal prohibition to prevent a disgruntled faction in an open-source project from taking off in a new direction. This is known as "forking," and one of the inalienable rights of the open-source community is the "right to fork." In theory, this leaves open-source projects in constant danger of fracturing into different groups, each developing a competing version of a program. In practice, things have not worked out that way in the two flagship projects; Apache and Linux have enjoyed persuasive and effective leadership. But the projects have also been held together by the economics of "network effects" – simply put, that the more people who use a product, the more valuable the product becomes. Network effects can be particularly powerful with a complex technological product like a software program. The more people, for example, who write applications programs that run on Apache or Linux, the more "locked in" they become, and thus less likely to pay the price in time, energy, and money that would be required to retool their software to run on some other server or operating system. It is precisely the same economic logic that worked to Microsoft's advantage in establishing its Windows operating system as a technology standard in the personal computer industry.

The difference between the two camps, though, is the insistence of the open-source community that technology standards remain open, rather than owned by any single company. The open-source philosophy comes from the Internet and its tradition that standards be in the public domain, from the early communications and routing software protocols like TCP/IP to the basic standard for transferring data over the Web, HTTP. The open-source community believes that not just the communications protocols, but anything that is "framework" software – software that is used by nearly everyone – should be open source. Apache, an HTTP Web server, is one such program, and they insist the same should be true of operating systems – hence Linux. That stance, of course, places the open-source community squarely at odds with the dominant software company, Microsoft, which has moved its Windows operating system beyond the PC desktop into the market for industrial-strength operating systems on big server computers used to power everything from corporate data systems to Internet commerce. But the open-source community believes that Internet technology, economics, and destiny are on their side. "Let's put it this way," Behlendorf said, "I don't think Microsoft's future as an operating system company is very bright."

Behlendorf dropped out of Berkeley at the end of his junior year in 1994 to become fully engaged in Internet computing and the open-source community. "I haven't regretted it," he said. After leaving *Wired* in 1995, Behlendorf was a founder and chief technology officer of Organic Online, a Web design and consulting firm, and in 1999 went on to help found Collab.Net, where he is a cofounder and chief technology officer.

Behlendorf is a pragmatist who speaks at length of the economic and business justification of open source, but he is also a believer. He believes that software is important not only as a business tool, but socially and politically. He compares the open-source movement broadly to the American Revolution. "It's all about control and decentralizing that control," he explained. "Programming is power, because code implements the policy of the creator. And the more knowledge you have in that environment – the programming environment – the more power you have."

Like many others, Behlendorf sees a future of pervasive digital networks in which software is the animating technology of commerce, communication, and education. He believes that it should be a democratic goal of society that everyone should have some control over the software than otherwise controls them

– to be able to tweak their software environment for privacy, security, fun, or whatever. That is, to be able to *program*. "Programming has to be at least accessible to everyone," Behlendorf said. "That doesn't mean you have to learn C or C++. But not being able to program is going to be like not being able to drive – lacking a fundamental skill in our society."

Craftsmen often grow fond of their tools. A workbench and its implements become a familiar, self-contained environment of satisfaction and achievement. The bond can be especially powerful among programmers as they become wedded to their computers and their software. A favorite machine, programming language, or operating system can shape the practitioner's point of view and mentality. Each language or operating system, for example, has strengths and weaknesses. Each offers the programmer certain freedoms and some restrictions, a set of rules that amount to a technical philosophy – a working culture, really. In extreme cases, the craftsmen-tool bond runs even deeper, becoming almost a way of life if not a religion. There is no more striking example than the conversion of Richard M. Stallman when he began working on the operating system of a Digital Equipment PDP-10 minicomputer in the distinctive setting of the MIT Artificial Intelligence Lab in the 1970s.

Stallman, who grew up in New York City, was something of a math prodigy and began dabbling with calculus at the age of eight. A few years later, one of his summer camp counselors happened to have a manual for the IBM 7094 mainframe, and Stallman picked it up and read it. He began to write some simple programs, like one to take a list of numbers, cube each one, and then add up the cubed numbers. He did not have access to a computer, but Stallman recalled, "I just wanted to write programs. I was fascinated." During high school, he took part in academic-enrichment programs that allowed bright students to get time on IBM mainframes, and Stallman was so adept that IBM hired him for a summer before he went to Harvard University. At Harvard, he got access to a time-shared computer in the applied math department. Because resources were limited, there was no permanent storage, so a person's work was deleted when he or she logged off the time-sharing machine. But Stallman soon discovered that if he signed on to two terminals at once, his programming files would not be deleted. It was a programming trick he performed for fun, to show he could do it. He never hogged storage space or terminal time. "I did not like rules," he recalled. "I still don't. It was an intellectual exercise expressing what I thought of rules." In software, Stallman – slight, shy, and cerebral – had found his favored medium for expressing himself and building things. "The only thing I ever built anything in is software. And with software, there are not the constraints of matter that you have with physical things. It's as if you could support a thousand-pound weight with a straw."

One day in 1971, at the end of his freshman year, Stallman wandered across Cambridge, Massachusetts, and into the MIT Artificial Intelligence Lab. He left the lab with a summer job as a systems programmer, and though Stallman would graduate from Harvard in 1974 with a bachelors degree in physics, his consuming passion would be building code at the MIT lab until he walked out in 1984. In his 13 years at the Artificial Intelligence Lab, Stallman did very little artificial intelligence programming. He worked mainly on the operating system for the PDP-10 machine in the lab. In the early 1970s, the MIT Artificial Intelligence Lab was a place where academic freedom, technological optimism, and an anti-establishment culture flourished. It was an open environment where everything was shared, nothing was owned, and rank and privilege did not exist. Working on the operating system, Stallman recalled, was like playing in a communal sandbox of software. "Anybody with ability was welcome to come in and try to improve the operating system. We built on each other's work. . . . I could start where they left off."

In the lab, Stallman also became intimately familiar with Lisp, the language of artificial intelligence. For Stallman, Lisp seemed an ideal programming language, imaginatively designed and without the inhibiting rules of other computing tongues. Lisp shunned the autocratic hierarchy of data definitions of other languages, and its simple syntax drew no distinction between data and programming instructions. Even today, Stallman describes the appeal of Lisp with heartfelt passion, speaking of "its elegance, its very simple syntax, its flexibility and its power."

Programmers had power at the MIT lab, a reality that inspired Stallman and had a kind of politicizing influence on him. "I was timid about standing up to authority," he said. "But the AI lab taught me to do that." Shortly after he arrived, Stallman noticed a heavy steel cylinder mounted on wheels. It had been used as a battering ram to break down the door of a professor who had locked a terminal in his office. "Anybody who had a terminal and wasn't using it had to cede it," Stallman explained. The programmers' rules were enforced on everyone, even an MIT professor trying to hog a terminal. "The hackers at the AI lab didn't allow that," he said.

By 1981, however, things began to change at the MIT lab. People departed to start companies and make money producing specialized Lisp machines that borrowed heavily from the programming work done at the MIT lab. But the companies added proprietary features, fought each other, and refused to share. His old friends had left the lab, and its special culture, in Stallman's view, had been destroyed by short-sighted corporate greed. In January 1984, Stallman resigned from MIT to pursue his seemingly quixotic mission of freeing the software world. "The conclusion I reached was that proprietary software is wrong," he said. "It is based on dividing people and keeping them helpless. I decided to fight it and try to bring about its downfall."

With long flowing hair, a beard, and green eyes that dance, Stallman sat in his mother's kitchen in New York one winter day in 2000, alternately tugging at snarled ends in his hair and pecking Lisp code on a small silver notebook computer while answering questions. He is an odd man, but also oddly charismatic, with a quirky sense of humor. In *The Hacker's Dictionary,* published in 1983, Stallman, who was a contributor, wrote of himself, "I was built at a laboratory in Manhattan around 1953, and moved to the MIT Artificial Intelligence Lab in 1971. . . . About a year ago I split up with the PDP-10 computer to which I was married for ten years. We still love each other, but the world is taking us in different directions." Friends and colleagues call him by his initials RMS, an homage to the way he logged onto the computer at MIT, which Stallman apparently prefers. "Richard Stallman is just my mundane name," he wrote in 1983.

When he left MIT in 1984, Stallman waged his war against the proprietary regime with his only weapon – writing software and distributing it free. He chose as his target of opportunity the Unix operating system, licensed by AT&T, and he called his project GNU, a playful recursive acronym for "Gnu's Not Unix." Stallman is widely acknowledged to be a great programmer. In 1990, he won the Association for Computing Machinery's Grace Hopper Award as the "outstanding young computer professional" of the year for his development of Emacs, an editing program that has become a standard among sophisticated programmers. An established operating system like Unix includes many ingredients – an array of development tools, libraries, and compilers – and in 1984 Stallman began the Olympian task of trying to match or surpass Unix in every category. His work would be shared with the world, and he invited others to join. Over the years, Stallman and some collaborators made real, even striking, progress in developing their Unix-compatible operating system. The GNU compiler and systems libraries were excellent, and GNU even included a chess game. After all, Unix included a chess game, so GNU would too. "That's something a computer should do – play chess," Stallman explained.

Yet Stallman had not gotten to the task of designing an operating system "kernel" – the program's core, closest to the machine, controlling the most basic operation. But in 1991, over in Finland, Linus Torvalds, a student at the University of Helsinki, was taking his own approach to the kernel problem. He had taken his school's first course on "C and Unix," and he was smitten by the Unix philosophy and the "clean and beautiful operating system" it had produced. Suitably inspired, Torvalds made a down payment on a $3,000 personal computer, and eventually wrote his own Unix-type kernel. Torvalds distributed his work on the Internet, and enlisted the help of interested programmers. They gathered on-line, slowly at first but in ever-increasing numbers. They were impressed with Torvalds' code and his light but sure touch as a project leader. They even took up a collection to pay for his computer, which he had purchased on a three-year installment plan. They weighed in with suggestions, refinements, and improvements. But a well-crafted kernel, though vital, is only one ingredient of an operating system. For the rest, Torvalds and his followers used the available free software, including the GNU compiler as well as software developed by the BSD project, a Unix variant developed at the University of California at Berkeley starting in 1979. The resulting operating system was called Linux, as if a contraction of "Linus's Unix."

The Linux project was founded just in time to ride two waves of technology – the rise of increasingly powerful and affordable personal computers and the Internet explosion. Suddenly, inexpensive machines could run more sophisticated software, and the Internet made it far easier for ambitious programmers to communicate and collaborate. Eric Raymond, the open-source essayist, observed that Torvalds was not really a software design wizard in a class with Stallman or James Gosling, the creator of Java, but that his gift was mainly in engineering a new model of software development. Torvalds, Raymond wrote, was "the first person who learned to play by the new rules that pervasive Internet access made possible." When asked, Torvalds replied that "my real talent has been a good combination of technical skills and communication." Later, he mused, "What made Linux special? Probably the fact that I wasn't very political." Torvalds does hold firm to the principle of controlling what happens to the Linux code, and demanding that all enhancements to the operating system be placed back into the project, into the public domain. "So I made people improve Linux," he said, "without getting too upset about the silly political agenda that the Free Software Foundation tries to push."

What seems silly to Linus Torvalds is the heart of the issue to Richard Stallman. There is a lot that irritates Stallman, but he is particularly chagrined when people assume he is part of the open-source community, as they often do. "That's like calling FDR a Republican," he huffed. The open source message is mainly one of efficiency — that its model will generate better software for users, even if some software suppliers will suffer. But Stallman's free software movement emphasizes the morality of its cause. "The principal aim of open source is not freedom, but success," Stallman said. "What a shallow, pointless goal." Stallman wages his campaign with words, which he deploys with the self-righteous certainty of a moralist and the insistent precision of an engineer. So, in his definition, free software and open source software are worlds apart, even though their practices mostly overlap.

Similarly, Stallman insists that the proper name for Linux is "GNU Linux," because of the contribution of the Free Software Foundation's GNU project. To call it simply Linux, he declared, is "unfair and unjust." In reply, Torvalds says he doesn't much care what the operating system is called. Still, Torvalds added that Stallman's argument for GNU Linux was "always fairly weak, in the sense that there's more code from BSD," the Berkeley Software Distribution, and other open-source projects than from the Free Software Foundation in Linux. Yet, once again, it appears that ideology, more than code, is what rankles Stallman. "There are millions of users that are using the system but have no idea of the philosophy that created it," he explained, noting forlornly that most Linux programmers regard the GNU contribution as merely a handful of tools. The open-source people, Stallman said, "give us credit for individual trees, but do not mention that we built the forest."

Stallman's free software campaign over the years has prodded many people to rethink how software is produced, and reconsider the implications of treating code as intellectual property to be owned by individual companies. But the reason freely distributed software now runs on millions of computers is because of open-source moderates like Linus Torvalds, a fair-haired, clean-shaven Finn, who now lives in Silicon Valley. To many, Torvalds is the acceptable face of hackerdom – practical, self-deprecating, and welcoming, right down to his choice of a cuddly, chubby cartoon penguin as the Linux mascot. Some of this, of course, is a marketing tactic – a modern "plain folks" appeal to his intended audience. He can be as caustic and dogmatic on technical issues as Stallman. In fact, Torvalds has said he hates the GNU Emacs editor – a program many others regard as an imaginative work of art – "with a passion," dismissing it as simply "horrible." Yet Torvalds is an attractive ambassador for open source software. Corporate audiences often tend to reject the ideological intensity of Stallman, who lives off the invested proceeds of a MacArthur Foundation fellowship and speaking fees, as off-putting and threatening. By contrast, Torvalds is a card-carrying member of the corporate world, working for Transmeta, writing software for a microchip maker.

Torvalds does not, he noted, have the "MIT radicalism background" of Stallman and others in the Free Software Foundation. "RMS," Torvalds said, "is the Buddhist monk that sets himself on fire to get people to wake up. He's *that* convinced about the need of open source. And that's great. But at the same time, that makes him a very hard person to interact with on any sane level, except as an icon. And it makes a lot of people really dislike him and everything he stands for. He's too inflexible, too religious. . . . I certainly am of the opinion that open source started working a lot better once it got away from the Free Software Foundation politics and values, and more people started thinking of it more as a tool than a religion. I'm definitely a pragmatist."

The companies lining up behind the open source banner are inspired by pragmatic self-interest, and none more so than IBM. The world's largest computer company embraced Apache in 1998, after its own Web server software failed to attract a following. The big move, though, came at the start of 2000, when IBM declared that it would throw its considerable weight behind Linux. It is sound business strategy for IBM to try to transform the operating system into a profitless commodity, thus undermining two of its leading rivals, Microsoft and Sun Microsystems. In the market for modern operating systems, outside its mainframe bailiwick, IBM is a loser. Its effort in personal computer software, OS/2, was handily crushed by Microsoft's Windows, and IBM's AIX flavor of Unix has become an also-ran behind Sun's Solaris. So to IBM, following the time-honored logic that "the enemy of my enemy is my friend," the pudgy little Linux penguin looks downright lovable.

But more significantly, IBM has also come to believe that Linux – and its open-source development model – will be a winner. IBM's principal business asset is its long-term relationships with corporate customers. If IBM leads them astray technologically by placing competitive considerations above the welfare of its corporate customers, they will abandon Big Blue. So IBM has a lot riding on Linux and open source. The IBM decision to bet on Linux was made at the top, but the strategy was shaped further down. A crucial convert was Irving Wladawsky-Berger, a Ph.D. physicist who is a trusted technology adviser to the company's top executives. A longtime IBM veteran, Wladawsky-Berger straddles the worlds of business and research. He carries the title of a corporate vice president, but he is not really on the corporate ladder. He personally briefed chairman Louis V. Gerstner Jr. and other top executives before IBM made its move in January 2000. His message was that Linux and open source appeared to be "another disruptive technology," like the rise of the microprocessor that caught IBM on its heels in the 1980s, and the Internet, which IBM moved quickly to assimilate in the 1990s. "I can't tell you exactly where it's going," he told them, "but it feels like another big thing."

Wladawsky-Berger was directly involved in prodding IBM to adjust to the previous two waves of disruptive technology. During the 1980s, he was vice president for development in the mainframe division, where he warned that low-cost microprocessor technology was destined to erode IBM's big-computer stronghold. Not only could smaller machines running microprocessors take over chores previously handled by IBM's giant machines, but the mainframes themselves would inevitably have to shift to mass-produced microprocessor technology instead of bipolar processors, which were essentially handcrafted. But Wladawsky-Berger concedes even he did not foresee how quickly the microprocessor revolution would take over, bringing massive losses and management upheaval at IBM. When the Internet took off in the 1990s, Wladawsky-Berger was named general manager of the Internet division, and his job was to make sure the company would not be caught flat-footed again. His role was to act as an evangelist and steward of IBM's Internet strategy, making sure that every one of the company's businesses was in step. During the 1990s, IBM nimbly positioned itself as supplier of Internet technology and expertise to companies large and small, and it prospered. Big Blue had learned its lesson: move early and exploit the next wave of technology, instead of trailing and being forced to defend the past.

The son of Jewish Eastern European immigrants, Irving Wladawsky-Berger was born and raised in Cuba. His hyphenated last name (pronounced Vla-daw-skee-Berger), he said, was "part of trying to be Cuban, strange as that may seem now." His Russian father, Julius, owned a five-and-dime convenience store in Havana, and he lost everything after Fidel Castro came to power in 1959. The family came to America in 1960 with little, though the teen-aged Wladawsky-Berger brought from Cuba a lifelong love of baseball. He is a New York Mets fan, having never forgiven George Steinbrenner, owner of the New York Yankees, for trading Reggie Jackson in 1982. Owing something to his Cuban upbringing, he also has a taste for Latin jazz – Poncho Sanchez, Tito Puente, Paquito D'Rivera, and others – and he is an avid consumer of the novels of Elmore Leonard. "I love the scenes of Miami lowlife," he noted.

In the United States, his family settled in Chicago, where they had relatives. Wladawsky-Berger quickly learned English and was an outstanding student. In the summer of 1962, before his freshman year at the University of Chicago, Wladawsky-Berger got a job at the university's new computer center, and he kept on working at the center while he was a university student. His early programming was mainly to help professors put scientific problems on an IBM 7090 mainframe, such as the mathematical-modeling program for a biologist that plotted how squid neurons and axons reacted to electrical impulses. He programmed mostly in an assembly language called FAP, and debugged by going through "octal dumps," a printout of the contents of the machine's memory in base-8 numeric notation.

"Programming was really fun," Wladawsky-Berger recalled. "Getting the computer to do what you wanted it to do, and getting the program to work, no matter how complex, was a real sense of accomplishment. You just felt this incredible feeling of power." He found the debugging to be "detective work," since he had to "track down precisely what was wrong." He would often stay up all night trying to run recalcitrant programs, submitting his code, then going out at midnight to eat at Unos or Dues (still-famous Chicago pizza places), or Taquerias, a Mexican restaurant south of downtown, then returning at 2 or 3 A.M., "hoping the job had run this time," Wladawsky-Berger recalled. "At 19 or 20, this felt like a really cool thing to do, which I guess qualifies me as a real nerd."

Small, bespectacled, with curly graying hair, Wladawsky-Berger is IBM's corporate nerd. In 1999, he began hearing more and more about Linux from researchers in the supercomputing field, from Internet programmers, and from his peers on a White House information technology advisory group composed of leading computer scientists. IBM was already backing Apache, and it had a small but growing corps of programmers participating in open source projects, including Linux. At the end of October, Samuel Palmisano, a senior vice president, had just returned from a global tour. The companies he visited told him that the preferred operating system of young Internet programmers was Linux. Palmisano, who was later named IBM's president, ordered up a "corporate assessment" on how IBM should respond to Linux.

On Saturday afternoon, October 30, 1999. Nick Bowen, a senior researcher at IBM, got a call at his home in Newtown, Connecticut. The caller was his boss, Paul Horn, the head of the Watson Labs. Bowen's assignment was to head an 11-person team to study Linux – the opportunities and the risks – and deliver recommendations. The Bowen report, presented to Palmisano on December 20, was a plan for undercutting the advantage enjoyed by Microsoft and Sun in operating systems used on data-serving computers. The goal would be to win the hearts and minds of perhaps the most influential audience in computing, the software developers who write the applications that bring the Web to life and make electronic commerce actually work. "Today, Microsoft and Sun dominate the application development seats," the report stated. "We recommend that IBM aggressively pursue a Linux-based application development platform and support Linux across all its products. Doing so would disrupt the Sun–Microsoft stranglehold." The Linux strategy, the report added, would give IBM a certain cutting-edge image as well, allowing it to compete with Sun and Microsoft for "mindshare in key software growth segments and in universities."

Palmisano liked what he read. After conferring with Gerstner, he issued an e-mail memo to the company's senior management team on January 7, 2000 that declared, "We will embrace Linux." Wladawsky-Berger was appointed IBM's Linux czar, and he proceeded to mingle in the open-source community. With his low-key professorial manner and his deep technical background, Wladawsky-Berger, the articulate IBM nerd, would seem an ideal emissary for the company that personifies corporate computing, as it embraces software's counterculture. "Linux," Wladawsky-Berger observed, "isn't so much an operating system as it is a culture."

Companies like IBM are not the only ones seeking an edge from the potentially disruptive force of open-source. A technology advisory group to the European Commission termed open source software "a great opportunity and an important resource" for the region. "European companies and developers are already a driving force in many open source projects," the group wrote in 2000 in its report, *Free Software/Open Source: Information Society Opportunities for Europe?* "If open source software is able to change the rules in the information technology industry, the companies and countries which better understand it and are more advanced in its use and knowledge will have a clear competitive advantage." And Erriki Liikanen, the Commissioner for Enterprise and Information Society, said that by focusing on open-source projects like Linux, "it may be possible to spark European creativity in this area and dramatically reduce our reliance on imports."

Just how far open source can go remains to be seen. New open-source ventures are bound to struggle as they seek to make a living in a business environment dominated by traditional commercial software companies. In May 2001, Eazel, the open-source startup populated with alumni from the Apple Macintosh team, closed its doors, unable to convince investors to continue supporting its unproven business model. Andy Hertzfeld, a founder of Eazel, which developed a graphic-icon user interface for Linux, emerged from the experience saddened but still enthusiastic about open source. "It's so much better for the users and developers that it will eventually prevail," he said. "It's just a matter of time." Hertzfeld walked away from Eazel impressed by the open-source development model, a "self-organizing meritocracy" in which "the cream rises to the top and first-class technical decisions usually emerge from the chaos." He was less impressed, though, with what he called "some of the hubris – open source, obviously, isn't the only way to make great software."

At the University of California at Berkeley during the late 1970s and early 1980s, Bill Joy was practicing the "open source" ethos long before the term was used. He was the principal designer of the Berkeley Unix, or BSD, and he distributed it freely. "In my mind, BSD was research," he recalled. "And if it was research, it should be published." An impressive creation, Berkeley Unix was much in demand at universities and research labs. Joy was the programmer who took the nascent Internet communications protocols, TCP/IP, and incorporated them, debugged and honed for performance, in Berkeley Unix, establishing the Internet's open protocols as the networking standard at universities instead of the proprietary software of DEC, IBM, or some other company.

Joy is at turns bemused, approving, and dismissive of the open-source move-ment. The Internet and the spread of inexpensive personal computers, Joy notes, has changed the context of things, making collaboration and rapid debugging easier. "But the truth is, great software comes from great program-mers, not from a large number of people slaving away," said Joy, a founder of Sun Microsystems in 1982. "Open source can be useful. It can speed things up. But it's not new, and it's not holy water." And he is skeptical of the wisdom of abandoning a business model that treats the programmer as a master craftsman or an artist. "Money flows to creators as a reward for creativity," Joy said. "I think it's too early to say that world is over. It works pretty well."

If the past is any guide, open source will be another tool for a profession of toolsmiths. The case for open source is it will help programmers build better software in some cases and do it faster – a power tool, perhaps, in the crafts-man's kit but not a magic wand. The Danish creator of C++ and student of philosophy, Bjarne Stroustrup, observed that "open source is a good idea that some people are trying to make the only idea, and that's always a mistake."

The world is not binary, even if it is more and more made of software, built bit by bit.

Afterword

PROGRAMMERS USE THE WORD "TOOLS" OVER AND OVER to describe both the software they create and the programming implements they use. To speak of tools is to adopt the language and the perspective of the craftsman and the engineer. Programmers labor in the field of "computer science," but it may be a misnomer. "It's an engineering discipline, not a scientific discipline," noted Fred Brooks, the software engineering expert. The two perspectives are quite different. The scientist builds in order to study, Brooks explains, while the engineer studies in order to build.

Many leading programmers fell into software from science or mathematics, after they found greater appeal in the practical tools of software than in the theory of other fields. Butler Lampson, whose research at Xerox PARC helped lay the foundation for the modern personal computer, described the conversion experience. During the 1960s, he was studying for his doctorate in physics at Berkeley, but one day he walked into the university's computer lab and never returned to physics. "I stumbled in there, and got sucked into computing," he recalled. And like so many physicists and mathematicians who lapsed into computing, Lampson speaks of the satisfactions of engineering – of making things with software, a building material that requires only intellect to assemble, not armies of workers or trucks and cranes to hoist. Programming, Lampson says, is free-range engineering. "Engineers get their kicks out of building things. That's the drug for just about anyone who has gotten hooked on computing."

As this book is being finished, the next evolutionary step in computing is coming into view. It has been dubbed "grid" computing, suggesting that intelligent computer power will someday be available to people like a utility, assisting people whenever and wherever it is needed – just plug in or tap in wirelessly, and the power of a supercomputer will be available. In many ways, the grid concept is remarkably familiar, echoing J. C. R. Licklider's vision of "man–machine symbiosis" and the information utility idea dating back to the

1960s. Like innovations from time-sharing to the Internet, grid computing has begun in the research labs of universities and governments.

Like every other important step in computing, the grid concept is starting to become practical because of continuing advances in processing power, network capacity, and software. In its grandest formulation, the grid will become a technology infrastructure for solving human problems, large and small, from curing diseases to playing games. The grid, though well in the future, could be a realization of the decades-old vision of allowing much of society to use computers to "augment" human intelligence.

The grid – if it happens – will depend crucially on software. The grid software can be thought of as a big step beyond something like Java, effectively making the Internet far more intelligent, more programmable. Ian Foster, a native of New Zealand who is now a senior scientist at the Argonne National Laboratory near Chicago, is one of the leaders in designing software for the grid. As is so often the case, he and a colleague, Carl Kesselman of the University of Southern California, began working on one programming problem years ago, and their work expanded. "The tools we were building had broader uses," Foster recalled. Such has been the story in software from FORTRAN to Unix to the Web – the programmer's vision expands as the tools prove increasingly useful.

The creativity in programming is of the engineering variety, part practical invention and part architectural design – some Thomas Edison and some Frank Lloyd Wright. In software, breakthroughs are not defined in terms of the grand theoretical insights of science and math, of $E = MC^2$-style epiphanies. Instead, ever since FORTRAN, programming innovation has been a matter of seeing a way to build something new or do it in a new way, and then doing it. In that, the software artist's craft is more like that of an architect or a civil engineer – a matter of designing and building, but with a difference. Today's big software programs are more complicated than anything else built by man.

The best programmers have the mental muscles for both conceptual thinking and procedural detail, and the uncanny ability to shift back and forth effortlessly, from high-level design to close-to-the-machine implementation. And for the people who are good at it, using those mental muscles is not only fun, but oddly compulsive. It may be difficult to explain to those outside the programming community, but Stanford's Donald Knuth, the author of *The Art of Computer Programming*, did a pretty good job. In his sixties, Knuth explained, he still feels the need to write programs every day. "I have to program because of the aesthetics of it," he said. "I love to see the way it fits together and sort of sings to you."

Notes

1: Introduction: The Rise of Software and the Programming Art

1   "no mercenary reasons." Author interview, October 10, 2000.

2   "a few lines of code." Author interview, November 18, 1999.

3   roots to the Babylonians. Martin Campbell-Kelly and William Aspray, *Computer: A History of the Information Machine* (Basic Books, 1996), p. 181.

3   a Persian scholar. Knuth, Donald E. *The Art of Computer Programming.* Volume 1: *Fundamental Algorithms,* (Addison-Wesley, 1997), pp. 1–2.

4   "every wire and every switch." interview, September 11, 2000.

4   "most exciting." From a 21-page personal history by Jean Bartik, dated February 10, 1995, p. 9.

4   "came over from England." Interview with Grace Hopper, ca. 1976, conducted by Christopher Evans, as part of the Charles Babbage Institute's oral history project. Transcript, p. 8.

5   "uniquely self-referential." Interview, January 5, 2001.

5   "By June 1949." Maurice Wilkes. *Memoirs of a Computer Pioneer* (MIT Press, 1985), p. 145.

6   first published use. The finding was made by Fred Shapiro, a librarian at the Yale Law School. "Origin of the Term Software: Evidence from the JSTOR Electronic Journal Archive," *IEEE Annals of the History of Computing,* 22 (2000): pp. 69–71.

6   "Early programming." *IEEE Annals of the History of Computing* vol. 6, no. 1 (January 1984), pp. 16–17.

6   "They took anyone." Interview, September 11, 2000.

7   "employing nearly 9 million." Stephen D. Hendrick and Ludovica Bruno, *The 2001 IDC Professional Developer Model,* June 2001. For 2000, the International Data Corporation figure was 8.739 million; the estimate for 2001 was 9.783 million.

7   "the new physical infrastructure." President's Information Technology Advisory Committee, *Information Technology Research: Investing in Our Future,* February 23, 1999.

7   "reasonable upper limit." Quoted in Knuth, p. 231.

9   "Some people are." Interview, June 19, 2000.

9   "There are a certain percentage." Interview, December 7, 2000.

2: FORTRAN: The Early "Turning Point"

12   "The plan for the new machine." Charles J. Bashe, et al., *IBM's Early Computers,* pp. 130–35.

12   "They each got a shot." From IBM documentary film on FORTRAN, 13 minutes, May 26, 1982.

13   "I figured there had." All of the quotes from Backus, unless otherwise noted, and much of the description of his background and work come from four interviews. Two were at his home in San Francisco, August 7, 2000 and September 29, 2000, and two by phone, April 27, 2001 and May 3, 2001.

13   "reached 10 people." Some IBM documents list 11 people, including Grace Mitchell. She worked on the reference manual and the distribution of FORTRAN I to customers. But she was not on the program development team, according to Backus and Ziller.

14   "We were the hackers," Interview, September 12, 2000.

15   the "turning point." "Programming Languages, Past, Present and Future," *IEEE looking.forward,* student newsletter, Fall 1996.

15   "ninety-five percent of the people." Interview, September 22, 2000.

15   "In the beginning." e–mail to author, September 11, 2000.

15   "I loved the fact." Interview, August 7, 2000.

19   "The 701 was $15,000." *IBM's Early Computers,* p. 162.

19   "He really understood." Interview, August 7, 2000.

19   "This, as you can imagine." All the quotes from Irving Ziller, unless otherwise noted, come from three interviews: September 12, 2000, November 2, 2000, and April 27, 2001.

20   "I said, John." IBM documentary file, 1982.

20   "We used to joke." Interview, September 18, 2000.

20   "Why can't programmers." Told by Dan McCracken, Interview, Marcy 13, 2001.

21   "The objective from the very early." Interview with Wheeler, conducted by William Aspray, May 14, 1987, as part of the Charles Babbage Institute's oral history project, p. 3.

22   "I felt that sooner or later." Interview with Grace Hopper, in Charles Babbage Institute oral history project, p. 14.

22   "A-0 compiler, and versions." Paul E. Ceruzzi, *A History of Modern Computing,* p. 85.

22   "Her compiler produced." Campbell-Kelly and Aspray, *Computer,* p. 187; and interviews with FORTRAN team.

23   "First true compiler." Ceruzzi, p. 86.

23   "We simply made up." *The History of Fortran I, II and III,* first published in 1979 and reprinted in the *IEEE Annals of the History of Computing,* vol. 20, no. 4 (1998), p. 70.

23   "In 1968, Edsger Dijkstra." *Communications of the ACM* vol. 11, no. 3 (March 1968), pp. 147–48.

24   "Since FORTRAN will." *Preliminary Report: Specifications for the IBM Mathematical FORmula TRANslating System, 11/10/54,* p. 2.

25   "We received almost no." *History of Fortran I, II and III,* p. 72.

26   "Six months, come back." IBM documentary film, 1982.

27   "You entered a world." Interview, May 3, 2001.

28   "surprising transformations" and "we would not." *History of Fortran I, II and III,* p. 74.

29   By way of example, the Fortran progamming example was helpfully provided by Dan McCracken, and the translations to a PC assembly language and then binary were done by a FORTRAN compiler supplied by Lahey Computer Systems Inc.

29   "You need the willingness." IBM documentary film, 1982.

30   "They were buried in it." Interview, April 28, 2001.

31   "It was a repetitive process." Interview, November 2, 2000.

32   The file was jokingly labeled." From interview with Ziller, September 12, 2000.

33   "It was a revelation." Interview, September 12, 2000.

33   "Maybe it's hindsight." Interview, April 27, 2001.

34   "Fortran, that impressed." Interview, December 6, 2000.

3: The Hard Lessons of the Sixties: From Exuberance to the Realities of COBOL and the IBM 360 Project

35   "There was tremendous resistance." Author Interview, September 18, 2000.

38   "Frankly, I was surprised." Quotes from John McCarthy, unless otherwise noted from interview at Stanford, December 6, 2000.

39   "The first genesis." Interviews with Corbato, conducted by Arthur L. Norberg, as part of Charles Babbage Institute's oral history project, April 18, 1989 and November 14, 1990, p. 12.

39   "Memorandum to P. M. Morse Proposing Time Sharing." January 1, 1959. Available on John McCarthy's Web site, www-formal.stanford.edu/jmc/

40   "The phrase 'artificial." Tape of lecture at the Computer Museum History Center, Moffett Field, Calif., on the "origins of artificial intelligence." 70 minutes. March 8, 2001.

40   "I propose to consider." "Computing Machinery and Intelligence," *Mind,* vol. LIX, no. 236 (October 1950).

41   "then your objections are religious." Tape of McCarthy lecture, March 8, 2001.

42	"Lisp had a freedom." Interview, May 15, 2001.
42	"called IPL 2." From interview with McCarthy, December 6, 2000.
43	"artificial intelligence labs are the places." Interview with Minsky, conducted by Arthur L. Norberg, as part of Charles Babbage Institute's oral history project, November 1, 1989, p. 5.
43	"In 1958, for example." H. A. Simon and A. Newell, "Heuristic Problem Solving: The Next Advance in Operations Research," *Operation Research*, January–February 1958, pp. 1–10.
44	"We thought everything was possible." Interview, May 16, 2001.
44	"A great part of the negative view." Interview, December 6, 2000.
44	"Human-level artificial intelligence." Tape of McCarthy lecture, March 8, 2001.
44	"the most vital tool." T. A. Wise, "IBM's $5,000,000,000 Gamble," *Fortune*, September, 1966, p. 119.
46	"the maximum use of simple English." Cited in Jean E. Sammet, "The Early History of Cobol," *History of Programming Languages,* Richard L. Wexelblat, ed. New York: Academic Press, 1981. Much of the chronology of Cobol's creation is included in the Sammet paper, pp. 199–242.
46	"It's hard to describe." All quotes from Jean Sammet, unless otherwise noted, and most of the description of her background and work came from an interview at her apartment building in Maryland, September 19, 2000.
48	"recommend a short-range approach." From "The Early History of Cobol," p. 202.
49	"Nobody could come up." Interview, December 1, 2000.
49	"We can't find a single." R. W. Bemer, "A View of the History of Cobol," *Honeywell Computer Journal*, vol. 5, no. 3 (1971), p. 132.
51	"Cobol did hierarchical data." Interview, February 6, 2001.
51	Repeatedly, she told. Persistence of the Hopper "bug" story explained in an article in Laurence Zuckerman, "If There's a Bug in the Etymology, You May Never Get It Out," by *The New York Times*, April 22, 2000. Section B, p. 11.
52	"Grace was modest in her way." Interview, March 7, 2001.
53	"cream of the existing crop." Interview, Mary 3, 2001. The Bald Peak conference description from interview with Sayre, and from details in *IBM's Early Computers*, pp. 368–71.
55	"A few months ago." Quoted in T. A. Wise, "The Rocky Road to the Marketplace," *Fortune*, October, 1966, p. 139.
56	"tirade on the need." Watts S. Humphrey, "Reflections on a Software Life," a paper done for the Software Engineering Institute, Carnegie-Mellon University, p. 10.

56   "His grandfather had been." All quotes from Humphrey, unless otherwise noted, and most of the description of his background and work, came from a phone interview with him, May 5, 2001.

58   "recognizing the overlap." Interview, June 26, 2001.

58   "second-system effect." Frederick P. Brooks, Jr., *The Mythical Man-Month* (1975; reprinted and updated, 1995). Reading, Mass.: Addison-Wesley Publishing, p. 55.

59   "A lot of the problems." Interview, September 14, 2000.

60      $500 million. Campbell-Kelly and Aspray, *Computer*, p. 199.

60   "The hours were long." Interview, September 14, 2000.

61   "There was a 'crisis' in." "Thoughts on Software Engineering," *Proceedings of the 11th* International Conference on Software Engineering, Pittsburgh May 15–18, 1989, p. 97.

62   In 1993, Fred Brooks. "Language Design as Design." The paper was presented at a conference on April 20–23, 1993, and included in a book on the proceedings, *History of Programming Languages II,* Thomas J. Bergin, Jr. and Richard G. Gibson, eds., (New York: ACM Press/Addison-Wesley, 1996), pp. 3–23.

4: Breaking Big Iron's Grip: Unix and C

63   "Ken Thompson could not wait." All quotes from Thompson, unless otherwise noted, come from an interview with him on September 22, 2000, at Bell Labs and from an e-mail interview on February 7, 2001, after he left Bell Labs to join Entrisphere, a telecommunications startup in Cupertino, Calif. Similarly, most of the descriptions of his background and his work come from those interviews. Useful background and detail were found in the available transcripts of a Unix oral history project that is underway and directed by Michael S. Mahoney, a historian at Princeton University.

63   "Except to threaten and harm us." From a rich trove of materials on the Berkeley Free Speech Movement, including numerous articles from *The Daily Californian,* on the Web. At www.fsm-a.org

66   "the chief alchemist." From a letter sent by McIlroy for Thompson's early retirement party at Bell Labs, at the end of 2000.

67   "We took off," posted on Thompson's Bell Labs Web site, Cm.bell-labs.com/cm/cs/who/ken/

67   "I am a programmer." From "Reflections on Trusting Trust," *Communication of the ACM,* vol. 27, no. 8 (August 1984), pp. 761–63.

69   "Dennis Ritchie took a different path." All the quotes from Ritchie, unless otherwise noted, and most of the description of his background and work come from two interviews at Bell Labs, September

22, 2000 and January 29, 2001, and two e-mail interviews, February 7, 2001 and February 15, 2001.

70   "Soon, both of them were mired in the Multics project." Descriptions of the MIT time-sharing projects come from interviews with Thompson, Ritchie and others. And good accounts of the details and culture of those MIT projects appear in Ceruzzi, *A History of Modern Computing,* Campbell-Kelly and Aspray, *Computer,* and Levy, *Hackers.*

71   "according to Brian Kernighan." All Kernighan quotes come from an interview on February 6, 2001.

72   "Perhaps the most important." D. M. Ritchie and K. Thompson, "The Unix Time-Sharing System." Paper presented at the Fourth ACM Symposium on Operating Systems Principles, IBM Thomas J. Watson Research Center, Yorktown Heights, NY, October 15–17, 1973.

73   Doug McIlroy, a Bell Labs scientist. His description of pushing for concept of connecting programs as if streams of data, and all quotes, unless otherwise noted, come from an e-mail interview, February 7, 2001.

73   "We should have some ways." Memo written by McIlroy, October 11, 1964, and Dennis Ritchie has posted it on his Bell Labs Web site, cm.bell-labs.com/cm/cs/who/dmr/mdmpipe.html.

75   "Yeah, I've seen editors." This comment and later one from George Coulouris, from e-mail interview, February 16, 2001.

75   "holds an early software patent." Kenneth L. Thompson, "US3568156: Text Matching Algorithm." Filed August 9, 1967; issued March 2, 1971.

78   "C traces its origins." The definitive description of the development of C and its predecessors BCPL and B is to be found in Dennis Ritchie, "The Development of the C Language." His paper is included in *History of Programming Languages II,* Thomas J. Bergin, Jr. and Richard G. Gibson, eds. New York: ACM Press/Addison-Wesley, 1996.

5: Programming for the Millions:
The BASIC Story from Dartmouth to Visual Basic

81   "For people who program." Kurtz quotes, unless otherwise noted, and some description of his background, came from two e-mail interviews, May 29, 2001 and June 25, 2001.

82   "one hour of valuable 704 time." Thomas E. Kurtz, "BASIC." A paper presented at a conference June 1–3, 1978, and included in *History of Programming Languages,* Richard L. Wexelblat, ed. New York: Academic Press, 1981, p. 516. Much of the description of the development,

background and features of the BASIC language comes from the
Kurtz paper.

82   "BASIC was an open city." All quotes from Cooper come from an
interview in Palo Alto, December 5, 2000, and two e-mail interviews,
May 29, 2001 and June 15, 2001. The description of his background
and his account of the role that his program, Ruby, played in the
development of Visual Basic also come from those interviews.

83   "Over the years." Quoted in the oral history, *Inside Out: Microsoft – In
Our Own Words*.

83   "We wondered." The Kurtz "BASIC" paper, p. 518.

83   "for example, Joseph Traub." The description of his going to Colum-
bia and his goals there, from interview January 16, 2001 and Joseph F.
Traub, "What Will Be the Intellectual Impact of Computers?" *Proceed-
ings: Educational Frontiers and the Cultural Tradition in the Contemporary
University*, vol. 11 (1982–83). From a General Education Seminar,
Columbia University.

84   "The people making policy." Interview, February 6, 2001.

84   "outrageous suggestion." Account of this conversation and of John
McCarthy's suggestion that Dartmouth ought to try time-sharing
come from John G. Kemeny and Thomas E. Kurtz, *Back to BASIC:
The History, Corruption and Future of the Language* (Reading, Mass:
Addison-Wesley, 1985), pp. 5–6.

84   "his family had emigrated." The most detailed account of Kemeny's
background that I found appears in Slater's *Portraits in Silicon*.

87   1968 article. John G. Kemeny and Thomas E. Kurtz, "Dartmouth
Time-Sharing," *Science*, vol. 162, no. 3850 (October 11, 1968), p. 226.

88   Dennis Allison, an instructor. All quotes from Allison and much of
the description of Tiny BASIC and that period come from an interview in Palo Alto, December 5, 2000. Good description of Tiny
BASIC and computing culture of that time are in Freiberger and
Swaine, *Fire in the Valley* and Levy, *Hackers*.

90   "The event that started everything." Quoted in the oral history, *Inside
Out: Microsoft – In Our Own Words*

90   Allen and Gates. The most detailed account of the Gates's early years
and founding Microsoft appear in Manes and Andrews' *Gates*.

90   "massive mistakes in that program," *Programmers at Work*, Susan Lammers, ed. (Redmond, Wash.: Microsoft Press, 1986), p. 76.

91   "Actually, making a BASIC." This quote from Gates, the quote in the
following paragraph and his comments about Visual Basic come from
an e-mail interview, June 22, 2001.

91   "time-sharing system in 1971." The most detailed account of all that the personal computer revolution, including Microsoft's original BASIC, owes to work done on Digital Equipment minicomputers is in Ceruzzi, *A History of Modern Computing.*

92   "Hobbyists like our BASIC." Quoted in the oral history, *Inside Out: Microsoft – In Our Own Words.*

93   "In 1976, Gordon Eubanks Jr." All quotes from Eubanks and description of his background from interview May 30, 2001.

95   "Meshing the two was." Some of the background on Visual Basic comes from interviews, October 13, 2000, with managers from Microsoft's Visual Basic and Visual Studio.Net units, Rob Copeland, Dave Mendlen and Chris Dias

97   "Shapiro recalled." e-mail Interview, June 11, 2001, revisiting a comment he made in 1998, while the Microsoft antitrust trial was in progress.

97   "We have absolutely taken advantage," Interview, June 11, 2001.

98   "That has been the same," at the Microsoft Professional Developers Conference in Orlando, Fla., July 12, 2000.

6: The European Influence: From Algol to Pascal to C++

100   "20 percent of IBM's workers." Cambell-Kelly and Aspray, *Computer,* p. 169.

100   "The Americans took an engineering perspective." Interview, February 28, 2001.

100   "BCPL stood for." Origins of C language and names come from Ritchie's "The Development of the C Language," and Bjarne Stroustrup, *The Design and Evolution of C++* (Boston: Addison-Wesley/Pearson Education, 1994), p. 64.

100   "It has long been my personal view," From the Oxford University Computing Laboratory web site, quoting Strachey as describing the lab's philosophy. Strachey, who died in 1975, taught at Oxford after he left Cambridge. Web address is web.comlab.ox.ac.uk/oucl/about/philosophy.html

100   C++, Stroustrup says. All quotes from Stroustrup, unless otherwise noted, and most of the description of his background and work come from an interview, January 31, 2001, at the AT&T Labs in New Jersey, and two e-mail interviews, February 7, 2001 and February 28, 2001.

104   C++ is clearly the work of an individual. The discussion of Stroustrup's philosophy is most clearly stated in *The Design and Evolution of C++*, pp. 23–25.

105   The story of Algol. The description of Algol, its background and development comes from various sources and interviews. But two of the most useful were N. Wirth, "Recollections About the Development of Pascal," ACM SIGPLAN Notices, vol. 28, no. 3 (March 1993), and Maurice W. Wilkes, *Computing Perspectives* (San Francisco: Morgan Kaufmann Publishers, 1995).

106   "Here is a language." A comment first made by Anthony Hoare in "Hints on Programming Language Design," a lecture delivered at the ACM symposium on the Principles of Programming Languages in Boston in 1973. The first publication was in "State of the Art Report," in *Computer Systems Reliability*, C. Bunyan, ed., vol. 20, pp. 505–34. Pergamon Infotech, 1974.

106   "Their interest was more theoretical." Wilkes, *Computing Perspectives*, p. 95.

106   Niklaus Wirth, a Swiss computer scientist. His account and quotes come from "Recollections About the Development of Pascal."

107   "Pascal forced people to think." Interview, February 27, 2001.

108   Simula was developed for studying. Background on the development of Simula comes from a few printed sources. Holmevik, Jan Rune. "The History of Simula," (Oslo, Norway: Institute for Studies in Research and Higher Education, 1995). An earlier version of the article was published in *IEEE Annals of the History of Computing*, vol. 16 (4) (1994), pp. 25–37. A copy of the current version was posted on James Gosling's Web site, at java.sun.com/people/jag/Simula History.html. Other background on Simula and comments from Kristen Nygaard came from articles published on his Web site including, "How Object Oriented Programming Started." His Web site address is www.ifi.uio.no/~kristen. Also from Nygaard, Kristen. "Basic Concepts in Object Oriented Programming," ACM Sigplan Notices vol. 21, no. 10 (October 1986).

110   "In 1973, for example, Donald Knuth." From Holmevik, "The History of Simula," citing a 1992 talk by Nygaard.

111   "He was a determined." e-mail Interview, March 2, 2001.

112   "It was absolutely inspired." e-mail Interview, February 23, 2001.

113   "very farsighted engineering." Interview, February 26, 2001.

7: A Computer of My Own: The Beginning of the PC Industry and the Story of Word

115   "I almost fainted with delight." All of the quotes from Simonyi and most of the background on his family and work came from three interviews at his home outside Seattle, October 10, 2000, October

11, 2000, and October 13, 2000, and three e-mail interviews, March 14, 2001, March 16, 2001, and March 30, 2001. An interview with Simonyi appears in Lammers, *Programmers at Work*, which I read as background before meeting with Simonyi.

119   "Charles was never a nerd." Interview, March 14, 2001.

119   "largest allocation of 'friend and family' shares." Reporting by Patrick McGeehan of the *New York Times* from investment banking sources at the time of the offering. The fact was not printed in the *Times*.

120   "Only Charles would build," Interview, January 5, 2001.

123   "dressed in a 'debugging suit.'" Description of Simonyi's debugging outfit and all quotes from Thacker come from an e-mail interview, March 29, 2001.

124   "In the early 1970s, Xerox PARC." The most detailed account of Xerox PARC during those years is in Michael Hiltzik, *Dealers of Lightning: Xerox PARC and the Dawn of the Computer Age* (New York: Harper Business, 1999).

126   "And Lampson, with a little Tom Sawyer." Description of the development of Bravo comes from previously cited interviews with Simonyi and Lampson, and Lampson's paper "Personal Distributed Computing: The Alto and Ethernet Software," which was published in *A History of Personal Workstations* (New York: ACM Press, 1988), pp. 293–335.

128   "They had made extensive studies." Their research on how book editors work is described in Hiltzik, *Dealers of Lightning.*

131   "We had heard about Simonyi." This quote and Gates's later quotes describing Simonyi from e-mail interview, June 22, 2001.

132   "Yet when it signed the crucial contract with IBM." The account of Microsoft's purchase of its initial operating system from Seattle Computer is described in Manes and Andrews, *Gates.*

136   "They rarely speak of business or politics." Comments from Dawkins from e-mail interview, March 22, 2001.

137   "Since the mid-1990s." Description of Intentional Programming comes from previously cited interviews with Simonyi, and two papers by him. "Intentional Programming — Innovation in the Legacy Age," presented at IFIP WG 2.1 meeting, June 4, 1996; and "The Death of Computer Languages, The Birth of Intentional Programming," Technical Report, MSR-TR-95-52, September 1995

8: Computing for the Masses:
The Long Road to "Gooey" and the Macintosh

139   "Andy Hertzfeld learned an embarrassing lesson." All Hertzfeld quotes, most of the details of his background, and his later description

of working on the Macintosh come from an interview in Palo Alto, December 7, 2000, and four e-mail interviews, April 2, 2001, April 3, 2001, April 11, 2001, and May 30, 2001.

142   "really began with J. C. R. Licklider." The description of his work and his quotes, unless otherwise noted, come from his paper "Man-Computer Symbiosis," *IRE Transactions on Human Factors in Electronics,* March 1960: 4–11, also reprinted in Goldberg, ed., *A History of Personal Workstations,* pp. 131–40. A concise account of Licklider's career appears in Campbell-Kelly and Aspray, *Computer,* pp. 212–14.

143   "I thought, 'This is going to revolutionize.'" From transcript of interview with Licklider on October 28, 1988, conducted by William Aspray and Arthur Norberg, as part of the Charles Babbage Institute's oral history project, p. 28.

144   his professional manifesto. "Augmenting Human Intellect: A Conceptual Framework." Summary Report, Stanford Research Institute. October 1962. Also available on the Web site of Doug Engelbart's Bootstrap Institute, www.bootstrap.org.

145   "The seeds of Engelbart's career." Engelbart described his work and his career in a paper titled, "The Augmented Knowledge Workshop," published in Goldberg, ed., *A History of the Personal Workstation,* pp. 187–232, and a question-and-answer session was also published; pp. 233–36. The remaining quotes from Engelbart come from that paper.

147   At Xerox PARC, the terms of human-computer. Ceruzzi, *A History of Modern Computing* has a concise summary of the Xerox PARC work and its debt to the research of Licklider and Engelbart, pp. 257–63. Hiltzik, *Dealers of Lightning,* is the most detailed account of Xerox PARC in the 1970s.

148   "What it could do was quite remarkable." Most of the description of Kay's computing background and his work at Xerox PARC came from his paper, "The Early History of Smalltalk," *ACM Sigplan Notices* vol. 28, no. 3 (March 1993), pp. 1–52. The paper was presented at a conference on April 20–23, 1993, and is included in the book of the proceedings, Thomas J. Bergin Jr. and Richard G. Gibson, eds., *History of Programming Languages II* (New York: ACM Press/Addison-Wesley, 1996).

151   "Kay replaced the machine-like architecture." Kay's description of Smalltalk and his quotes on its legacy come from e-mail interviews, April 2, 2001 and April 4, 1001.

152   Lampson recalled the lack of a user interface. Interview, January 5, 2001.

153   "wow moments," Interview, March 14, 2001.

153   and understood by a few people, notably Jef Raskin. And Raskin has pointed this out himself, taking exception to accounts of the Macin-

tosh history that he felt gave Steve Jobs too much credit. His essay "Holes in History," in an ACM publication, *interactions*, July 1994, pp. 11–16, covers that ground. The most detailed single account of the Macintosh project is Steven Levy, *Insanely Great: The Life and Times of Macintosh, the Computer that Changed Everything* (New York: Penguin, 1994).

155   "longtime friend of Guy Tribble." The account of Tribble's friendship with Bill Atkinson and all quotes from Tribble from an interview in Palo Alto, September 26, 2000.

155   "It was one-hundred percent." All quotes from Capps and description of his work on the Macintosh come from an interview in Seattle, October 10, 2000, an interview in San Carlos, California, December 8, 2000, and an e-mail interview, April 2, 2001.

9: Programming for Everyman: Just Let the Users Do It

161   "In 1959, IBM assigned." Bashe et al *IBM's Early Computers*, pp. 474–80.

161   Don Chamberlin grew up in Campbell. All of the Chamberlin quotes, unless otherwise noted, and most details of his background and the discussion of his work comes from an interview in San Jose, December 8, 2000, and two e-mail interviews, April 17, 2001 and April 23, 2001.

162   In 1956, it cost $10,000 for a megabyte. From a paper by Chamberlin entitled, "A Brief History of Data" that he presented in a talk at the University of California at Davis, at the Department of Computer Science Colloquium, May 25, 2000.

162   In his 1973 paper. Charles W. Bachman, "The Programmer as Navigator," *Communications of the ACM* vol. 16, no. 11 (1973), pp. 635–58.

163   Yet Chamberlin took it seriously. Description of Codd's visit to Yorktown Heights and Chamberlin's comments on that visit come from "The 1995 SQL Reunion: People, Projects, and Politics." Edited by Paul McJones. SRC Technical Note, 1997–018. (The paper is a transcript of a conference held May 28–30, 1995 in Pacific Grove, California. The proceedings were published by Digital Equipment's Systems Research Center.)

163   "Codd's ideas, in fact." E. F. Codd, "A Relational Model of Data for Large Shared Data Banks," *Communications of the ACM* vol. 13, no. 6 (June 1970), pp. 377–87.

164   "Jim Gray was working at Yorktown." All the Gray quotes and most of the description of his background and work come from an interview in San Francisco, September 28, 2000, and an e-mail interview, April 14, 2000.

166   "Point of view is worth 80 IQ points." This is one of those quotes that pops up all the time when people are referring to innovation and research. But in a quick Web search using Google, I noticed the IQ numbers ranged from 50 to 100 in the quote attributed to Kay and "context" and "perspective" are frequently substituted for "point of view." So I e-mailed Alan Kay and asked what the original quote was, and he replied on April 19, 2001.

167   saying how useful he found a paper published in 1974. D. D. Chamberlin and R. F. Boyce, "SEQUEL: A Structured English Query Language," *Proceedings of the ACM SIGMOD on Data Description, Access and Control.* Ann Arbor, Michigan, May 1974, pp. 249–64.

168   "Bricklin's father, Baruch." All Bricklin quotes and most of the description of his background and work come from a phone interview, April 17, 2001. A profile of Bricklin that I read before interviewing him appears in Slater, *Portraits in Silicon.*

169   "Frankston's father, Benjamin." All the Frankston quotes and most of the description of his background and work came from an interview at his home in Cambridge, Massachusetts, January 13, 2001, and an e-mail interview, April 13, 2001.

171   "We had no accountants." Interview, January 5, 2001.

173   The most eloquent was Benjamin Rosen. A scanned copy of Rosen's report in 1979, "VisiCalc: Breaking the Personal Computer Software Bottleneck," can be found on Dan Bricklin's Web site, www.bricklin.com. The site also includes Bricklin's account of the creation of VisiCalc, with some pictures from those days.

174   "If you showed a fancy piece of software." Blasgen quote from "The 1995 SQL Reunion" transcript, p. 34.

174   "If you say 'HTML." Interview, September 26, 2000.

175   "Tim Berners-Lee, as it turned out." All of the Berners-Lee quotes, unless otherwise noted, and some of the background comes from two interviews, on December 13, 1995 and September 16, 1999. The most detailed account of his early working career and the evolution of his work and thinking that led to the Web appears in his 1999 book, written with Mark Fischetti, *Weaving the Web: The Original Design and Ultimate Destiny of the World Wide Web by Its Creator* (San Francisco: HarperSanFrancisco, 1999).

178 "It's really hard to demo the World Wide Web." Interview, September 26, 2000.

10: Java: The Messy Birth of a New Language

181 "This had disaster written all over it." All Gosling quotes and most of the description of his background and work come from interviews in Mountain View, California, September 26, 2000 and September 28, 2000, a phone interview, January 12, 2001, and an e-mail interview, June 29, 2000. He also wrote a Sun Microsystems "white paper," which is a good description of his language, "Java: An Overview," February 1995.

181 John Gage has the resume. A good background profile of Gage was written by Spencer Reiss, "Power to the People," *Wired*, December 1996

182 "Later, of course." Interview, January 12, 2001.

183 "James was uniquely qualified." Interview, January 11, 2001.

186 "But sure." Interview, January 16, 2001.

186 "He was known to be the best programmer." Interview, January 11, 2001.

189 "Looking back it sounds silly." Quote and Sheridan's description of atmosphere of the Java project in its early days from interview, November 15, 2000, and e-mail interviews, January 9, 2001 and June 29, 2001.

189 "the project was code named Green." "Behind the Green Door: Deep Thoughts on Business Opportunities in Comsumer Electronics." Presented by the Members of the Green Team, August 23, 1991 (Gosling gave me a copy of the paper.)

191 "came during a Doobie Brothers concert." That and other details were first printed in an article by David Bank, "The Java Saga," *Wired*, December 1995. I read David Bank's article, among others, before interviewing Gosling.

192 "A conversation with Joy." All Joy quotes and his account of working on Java come from an interview in New York, January 10, 1001, and e-mail interviews, October 9, 2000 and January 15, 2001.

193 "The genius of Java." All Schmidt quotes and his account of the development of the business strategy for Java come from an interview, January 15, 2001.

194 "C++ is an M-16." Interview, January 26, 2001.

196 "But I wasn't that impressed." All Steele quotes and his description of Java meetings come from an interview, January 19, 2001.

199   "We needed a way to make the Web programmable." All Andreessen quotes come from an interview, January 23, 2001.

201   "We think the time is right for another language." Interview, October 11, 2000.

201   "They're hard-core Java programmers." Interview, July 10, 2000.

11: There Has to Be a Better Way: Apple and the Open Source Movement

203   "We traded software patches." All Behlendorf quotes and most of the description of his background and work come from interviews in San Francisco, August 8, 2000 and September 26, 2000, and e-mail interviews, October 3, 2000, May 29, 2001, and May 31, 2001.

204   "We decided to take the code." All Terbush quotes and his description of the Apache project come from an interview, December 5, 2000, and an e-mail interview, May 31, 2001.

204   "In the spring of 2001, Apache ran on more than 60 percent." From "The Netcraft Web Server Survey," results for April 2001, reported in May 2001. Netcraft surveys Web server software use on Internet connected computers. Its surveys can be found on its Web site, www.netcraft.com

205   "In his essay." Eric S. Raymond, "The Magic Cauldron," included in his book, *The Cathedral and the Bazaar* (Sebastopol, Calif.: O'Reilly & Associates, 1999), pp. 137–94.

210   "I just wanted to write programs." All Stallman quotes and much of the description of his background and work come from interview in New York, December 23, 2000. Levy, *Hackers* has the most detailed account of Stallman's days at the MIT Artificial Intelligence Lab, and the beginning of his Free Software movement.

213   "clean and beautiful operating system." Linus Torvalds and David Diamond, *Just for Fun: The Story of an Accidental Revolutionary* (New York: Harper Business, 2001), p. 54 (from an uncorrected proof of the book) His book is the most detailed account of his background, and of the creation and development of Linux.

213   "the first person who learned to play by the new rules," Raymond, *The Cathedral and the Bazaar*, p. 63.

213   "my real talent has been a combination of technical skills and communication." Quotes from Torvalds, unless otherwise noted, come from e-mail interview, 6/3/01.

215   "with a passion." Linus Torvalds, "The Linux Edge," in *Open Sources: Voices from the Open Source Revolution,* Chris DiBona, Sam Ockman,

and Mark Stone, eds. (Sebastopol, Calif.: O'Reilly & Associates, 1999), p. 107.

216   "convert was Irving Wladawsky-Berger." Most of the Wladawsky-Berger quotes and the description of his education and programming experience come from an interview in New York, March 28, 2001, and e-mail interviews, July 3, 2000, August 2, 2000, and May 30, 2001. Most of the early personal background on him comes from a profile I wrote on Wladawsky-Berger, "Well, Somebody's Got to Reinvent the IBM Mainframe," *New York Times*, September 12, 1993, Sunday business section, p. 8.

218   "On Saturday afternoon October 30, 1999, Nick Bowen." The account of the IBM study of Linux and quotes from the report and executive e-mails come reporting for two stories I did for the *New York Times* on the subject. "IBM to Use Linux System in Software for Internet," January 10, 2000, Section C, p. 1, and "A Mainstream Giant Goes Counterculture: IBM's Embrace of Linux a Bet That It Is the Software of the Future," March 20, 2000, Section C, p. 1.

219   "a great opportunity and an important resource." From a report, *Free Software/Open Source: Information Society Opportunities for Europe?* By the working group on Libre Software (the group was formed at the request of the Information Society Directorate General of the European Commission). April 2000, Version 1.2. The paper is available on the Web at eu.conecta.ita.

219   "It's so much better for the users and developers." e-mail interview, May 30, 2001.

220   "Joy is at turns bemused, approving." Joy's comments come from interviews, August 6, 2000 and September 26, 2000.

220   "open source is a good idea that some people." Interview, January 31, 2001.

References

BOOKS

Bashe, Charles J., Lyle R. Johnson, John H. Palmer, and Emerson W. Pugh. *IBM's Early Computers*. Cambridge, Mass.: MIT Press, 1986.

Berners-Lee, Tim with Mark Fischetti. *Weaving the Web: The Original Design and Ultimate Destiny of the World Wide Web by Its Creator.* San Francisco: HarperSanFrancisco, 1999.

Bernstein, Jeremy. *The Analytical Engine: Computers – Past, Present, and Future*. New York: Random House, 1963.

Brooks, Frederick P., Jr. *The Mythical Man-Month*. Reading, Mass.: Addison-Wesley Publishing, 1975 (reprinted and updated 1995)..

Campbell-Kelly, Martin, and William Aspray. *Computer: A History of the Information Machine*. New York: Basic Books, 1996.

Ceruzzi, Paul E. *A History of Modern Computing*. Cambridge, Mass.: MIT Press, 1998.

Cusumano, Michael A. and Richard W. Selby. *Microsoft Secrets: How the World's Most Powerful Software Company Creates Technology, Shapes Markets and Manages People.* New York: Free Press, 1995.

DiBona, Chris, Sam Ockman, and Mark Stone, eds. *Open Sources: Voices from the Open Source Revolution*. Sebastopol, Calif.: O'Reilly & Associates, 1999.

Freiberger, Paul, and Michael Swaine. *Fire in the Valley: The Making of the Personal Computer.* New York: McGraw-Hill, 1984 (updated edition 2000).

Goldberg, Adele, ed. *A History of Personal Workstations*. New York: ACM Press, 1988.

Hillis, W. Daniel. *The Pattern on the Stone: The Simple Ideas that Make Computers Work.* New York: Basic Books, 1998.

Hiltzik, Michael. *Dealers of Lightning: Xerox PARC and the Dawn of the Computer Age.* New York: Harper Business, 1999.

Jerome, Kelli, and Marlee Anderson, eds. *Inside Out: Microsoft – In Our Own Words.* New York: Warner Books, 2000.

Kemeny, John G. and Thomas E. Kurtz. *Back to BASIC: The History, Corruption and Future of the Language*. Reading, Mass: Addison-Wesley, 1985.

Knuth, Donald E. *The Art of Computer Programming: Volume 1 Fundamental Algorithms.* Reading, Mass.: Addison-Wesley, 1997 (third edition).

Lammers, Susan, ed. *Programmers at Work.* Redmond, Wash.: Microsoft Press, 1986.

Levy, Steven. *Hackers: Heroes of the Computer Revolution.* New York: Doubleday, 1984.

Levy, Steven. *Insanely Great: The Life and Times of Macintosh, the Computer that Changed Everything.* New York: Penguin, 1994.

Manes, Stephen, and Paul Andrews. *Gates: How Microsoft's Mogul Reinvented an Industry — and Made Himself the Richest Man in America.* New York: Doubleday, 1993.

McCartney, Scott. *ENIAC: The Triumphs and Tragedies of the World's First Computer.* New York: Walker & Company, 1999.

Pugh, Emerson, and Lyle R. Johnson and John H. Palmer. *IBM's 360 and Early 370 Systems.* Cambridge, Mass.: MIT Press, 1991.

Raymond, Eric S. *The Cathedral and the Bazaar: Musings on Linux and Open Source by an Accidental Revolutionary.* Sebastopol, Calif.: O'Reilly & Associates, 1999.

Slater, Robert. *Portraits in Silicon.* Cambridge, Mass.: MIT Press, 1987.

Stroustrup, Bjarne. *The Design and Evolution of C++.* Boston: Addison-Wesley/Pearson Education, 1994.

Torvalds, Linus, and David Diamond. *Just for Fun: The Story of an Accidental Revolutionary.* New York: Harper Business, 2001.

Wilkes, Maurice V., and David J. Wheeler and Stanley Gill. *The Preparation of Programs for an Electronic Digital Computer: With Special Reference to the EDSAC and the Use of a Library of Subroutines.* Reading, Mass: Addison-Wesley, 1951.

Wilkes, Maurice. *Memoirs of a Computer Pioneer.* Cambridge, Mass.: MIT Press, 1985.

Wilkes, Maurice V. *Computing Perspectives.* San Francisco: Morgan Kaufmann, 1995.

Weizenbaum, Joseph. *Computer Power and Human Reason: From Judgment to Calculation.* W.H. Freeman & Company, 1976.

ARTICLES AND PAPERS

Bachman, Charles W. "The Programmer as Navigator." *Communications of the ACM,* vol. 16, no. 11, (1973), pp. 635–58.

Backus, John. *The History of Fortran I, II and III.* A paper originally published in the *Annals of the History of Computing* vol. 1, no. 1 (July 1979). It was reprinted in the *IEEE Annals of the History of Computing* vol. 20, no. 4 (1998), pp. 68–78.

Bank, David. "The Java Saga." *Wired,* December 1995.

Brooks, Frederick P. Jr. "Language Design as Design." The paper was presented at a conference on April 20–23, 1993, and included in a book on the proceedings, *History of Programming Languages II,* Thomas J. Bergin, Jr. and Richard G. Gibson, eds. New York: ACM Press/Addison-Wesley, 1996, pp. 3–23.

Bush, Vannevar. "As We May Think." *The Atlantic Monthly,* July 1945. Pp. 101–8.

Codd, E.F. "A Relational Model of Data for Large Shared Data Banks." *Communications of the ACM* vol. 13, no. 6 (June 1970), pp. 377–87.

Chamberlin D. D. and R. F. Boyce. "SEQUEL: A Structured English Query Language." *Proceedings of the ACM SIGMOD on Data Description, Access and Control.* Ann Arbor, Mich. (May 1974), pp. 249–64.

Dijkstra, Edsger W. "Go To Statement Considered Harmful." *Communications of the ACM* vol. 11, no. 3 (March 1968), pp. 147–48.

"Preliminary Report: Specifications for The IBM Mathematical FORmula TRANslation System." Programming Research Group, Applied Science Division, International Business Machines Corporation, November 11, 1954.

Holmevik, Jan Rune. "The History of Simula." Oslo, Norway: Institute for Studies in Research and Higher Education, 1995. An earlier version of the article was published in *IEEE Annals of the History of Computing* 16 (4) (1994), pp. 25–37. A copy of the current version was posted on James Gosling's Web site, at java.sun.com/people/jag/SimulaHistory.html.

Kay, Alan. "The Early History of Smalltalk." *ACM Sigplan Notices* vol. 28, no. 3 (March 1993), pp. 1–52. The paper was presented at a conference on April 20–23, 1993, and is included in the book of the proceedings, *History of Programming Languages II,* Thomas J. Bergin Jr. and Richard G. Gibson, eds. New York: ACM Press/Addison-Wesley, 1996.

Kemeny, John G. and Thomas E. Kurtz. "Dartmouth Time-Sharing." *Science* 1968. vol. 162, no. 3850 (October 11, 1968).

Lee, J. A. N. "Programming Languages, Past, Present and Future." *IEEE looking forward,* Student Newsletter, Fall 1996.

Lee, J. A. N., and H. S. Tropp, eds. "Special Issue: Fortran's Twenty-Fifth Anniversary." Vol. 6, No. 1, January 1984.

Licklider, J. C. R. "Man–Computer Symbiosis." *IRE Transactions on Human Factors in Electronics,* March 1960, pp. 4–11; also reprinted in Goldberg ed. *A History of Personal Workstations,* pp. 131–140.

Ritchie, D. M. and K. Thompson. "The Unix Time-Sharing System." Paper presented at the Fourth ACM Symposium on Operating Systems Principles, IBM Thomas J. Watson Research Center, Yorktown Heights, NY, October 15–17, 1973.

Ritchie, Dennis M. "The Development of the C Language." The paper was presented at a conference on April 20–23, 1993, and included in a book of the proceedings, *History of Programming Languages II,* Thomas J. Bergin, Jr. and Richard G. Gibson, eds. New York: ACM Press/Addison-Wesley, 1996.

Sammet, Jean E. "The Early History of Cobol." A paper presented at a conference on June 1–3, 1978, and included in a book of the proceedings, *History of Programming Languages,* Richard L. Wexelblat, ed. New York: Academic Press, 1981.

Thompson, Kenneth. "Reflections on Trusting Trust." *Communication of the ACM* vol. 27, no. 8 (August 1984), pp. 761–63.

Traub, Jospeph F. "What Will Be the Intellectual Impact of Computers?" *Proceedings: Educational Frontiers and the Cultural Tradition in the Contemporary University.* Vol. 11, 1982–83. From a General Education Seminar, Columbia University.

ORAL HISTORY TRANSCRIPTS

(All from the oral history project of the Charles Babbage Institute at the University of Minnesota.)

Corbato, Fernando J. Interviewed by Arthur L. Norberg, April 18, 1989 and November 14, 1990.

Hopper, Grace. Interviewed by Christopher Evans, ca. 1976.

Licklider, J.C.R. Interviewed by conducted by William Aspray and Arthur Norberg, October 28, 1988.

Minsky, Marvin L. Interviewed by Arthur L. Norberg, November 1, 1989.

Wheeler, David J. Interviewed by William Aspray, May 14, 1987.

VIDEOTAPES

Interviews with members of Fortran team. IBM documentary film, from IBM Archives. Produced by Bright Star Films. 13 minutes. May 26, 1982.

McCarthy, John. Tape of lecture at the Computer Museum History Center, Moffett Field, Calif., on the "origins of artificial intelligence." 70 minutes. March 8, 2001.

Allen, Fran. Tape of lecture at the Computer Museum History Center, Moffett Field, Calif., on the government's Stretch/Harvest computer. 114 minutes. November 8, 2000.

Index

Gage, John, 181–82
Galler, Bernard A., 61
Gates, Bill, 90–93, 95, 96, 116, 131–33
Gateway, 134
General Electric (GE), 70, 87
Gerstner, Louis V., Jr., 216
Gier computer, 120, 122
Gill, Stanley, 21
"Gnu's Not Unix" (GNU), 212–13
Goldberg, Adele, 152
Goldberg, Richard, 14, 26
Goldstine, Herman, 7
Gomery, Ralph, 165
Gorn, Saul, 45
Gosling, David, 184
Gosling, James, 8, 181–88, 190–202, 213
Gosling, Joyce, 184
Gramsci, Antonio, 118
graphical user interface (GUI), 134, 136,
 147
Gray, Jim, 15, 164–66
GreenTalk, 189–90
grid computing, 221–22
Guide to FORTRAN Programming, A
 (McCracken), 33
Gupta, Satish, 186
Gypsy interface, 128

Hacker's Dictionary, The, 212
Haibt, Lois, 6, 26–27, 31
hardware, 6
Harvard Mark I computer, 4
Harvard Mark II computer, 51
Harvest supercomputer, 36
Hawes, Mary, 45
Heckel, Paul, 129–30
Hejlsberg, Anders, 201
Hepburn, Katherine, 17–18
Herrick, Harlan, 19–20, 23, 25
Hertzfeld, Andy, 139–42, 154–55, 219
 as "Software Wizard," 142
Hewlett-Packard, 88
Hoare, Sir Anthony, 106
Hoff, Ted, 116
Homebrew Computer Club, 92
Hopper, Grace, 4, 8, 25, 46, 47, 49
 and "automatic programming," 22–23
 and development of COBOL, 51–53
 as "grandmother" of COBOL, 52

Horn, Bruce, 156
Horn, Paul, 218
HTML (Hypertext Markup Language),
 160, 177, 178–79
HTTP (Hypertext Transfer Protocol), 175,
 177, 178
Humphrey, Watts, 55–60, 61–62
"Hungarian," 125
Hurd, Cuthbert, 12, 19

IBM, 11–12, 17, 132, 134, 149, 153, 206,
 215–17, 219
 "Action Program," 53
 "alpha"/"beta"/"gamma" software, 53
 and the Blue Letters, 57
 and Deep Blue, 38, 75
 Defense Calculator, 11–12, 16
 early computers, 12–13
 and FORTRAN, 32–33, 35–37, 45
 1401 computer, 160–61
 and the "full circle of computing," 65
 hardware/software duality, 59–60
 and Linux, 215–16
 magnetic hard disk development, 162
 and OS/360 operating system, 54–55,
  57–60, 62, 65, 111
 resistance to FORTRAN, 35
 701 Defense Calculator, 18–19
 software development, 53–56
 System R project, 163–65, 168
information superhighway, 191–92
Ingalls, Dan, 152
Institute for Advanced Study (Princeton), 1
Intel Corporation, 116, 189
International Business Machines. See IBM
Internet, 183, 192–93, 203, 213. See also
 Java programming language
 and public domain, 209
 and software engineering, 205
IPL 2 list–processing language, 42
iPlanet, 204

Java programming language, 8, 43, 109,
 152, 159, 183, 201–2, 222
 design concepts, 193–96
 development, 188–93
 and Internet programming, 196–98
Jobs, Steve, 116, 140, 156, 188
Johnniac computer, 42